Penguin Handbooks

GOOD FOOD FROM YOUR FREEZER

Helge Rubinstein and Sheila Bush, both established cookery writers and authors of several popular cookbooks, share a love of cooking and good food. This, combined with practicality, enthusiasm and expertise, has led to their collaboration on such outstanding cookery books as the first, highly successful, *The Penguin Freezer Cookbook* and *Ices Galore* (to be published in Penguins).

Helge Rubinstein combines running homes in London and Oxfordshire with writing. Her most recent cookbook was *The Chocolate Book*, also published by Penguin, and her interest in chocolate has led her to develop a chocolate chip cookie business, Ben's Cookies, named after her youngest son. Formerly a marriage and family counsellor, she also writes on marriage and personal problems for two magazines, is co-author of the forthcoming *Fitness at 40*, and is the editor of *The Oxford Book of Marriage*, in preparation.

Sheila Bush began her professional life in publishing, later becoming a freelance writer and editor. She lives partly in London and partly in Sussex. She has written several books of general interest, and her cookbooks include *Italian Cookery* and *Recipes from a Château in Champagne*.

Good Food
From Your Freezer

****

Helge Rubinstein and
Sheila Bush

Penguin Books

Penguin Books Ltd, Harmondsworth, Middlesex, England
Viking Penguin Inc., 40 West 23rd Street, New York, New York 10010, U.S.A.
Penguin Books Australia Ltd, Ringwood, Victoria, Australia
Penguin Books Canada Ltd, 2801 John Street, Markham, Ontario, Canada L3R 1B4
Penguin Books (N.Z.) Ltd, 182–190 Wairau Road, Auckland 10, New Zealand

First published 1986

Typeset in 10/13pt Linotron Ehrhardt by
Rowland Phototypesetting Ltd
Bury St Edmunds, Suffolk
Made and printed in Great Britain by
Cox & Wyman Ltd, Reading

Contents

****

Introduction

****

In the first *Penguin Freezer Cookbook* we set out to show that, treated sensibly and with imagination, a freezer can greatly enhance your cooking, as well as making it more economical, by enabling you to use to best advantage fresh foods in season. In the present book we aim to supplement the first by concentrating on foods that are not seasonal: either fresh foods which can be frozen at home, before or after being cooked, or foods that are most easily available, and are often at their best, in commercially frozen form.

Since the first book was published in 1973, the number of homes with a freezer has increased by at least a million every year, so that now, some thirteen years later, over half the households in the UK (61 per cent, to be precise) own a freezer. And the range of frozen goods, both raw materials and ready-cooked dishes, has grown proportionately, in quantity as well as in quality.

Although saving money may not necessarily be the main aim, freezing can still achieve considerable economies, in terms of both money and time (which often comes to the same thing). Indeed, financial journalists regularly advise that stocking up the freezer at certain times of the year is a better investment than putting your money into a building society; and it has been claimed that up to 25 per cent of a family's food costs can be saved by well-judged buying and bulk-cooking and by cutting down on shopping trips.

There are many dishes where the basic recipe can be prepared as easily, and almost as speedily, in large quantities as in small ones, frozen in convenient amounts, and then quickly turned into delicious meals at short notice. In this way you can arm yourself for the days when there is no time to cook, or when you have guests, invited or uninvited.

Our hope is that this book will help readers to enjoy cooking more, by encouraging them both to use the freezer for a wider range of foods, and to cook less often from a sense of duty and more for sheer pleasure.

Choosing a Freezer

****

The greatest change in freezer appliances during the last decade has been the move towards combined refrigerator/freezers: while the ownership of freezers has gone up nearly five times, that of fridge-freezers has increased about thirty times. Perhaps this trend springs partly from the fact that shopping habits are changing. People no longer tend to buy food in very large quantities for freezing, and butchers, for instance, instead of selling whole or half carcasses, are now cutting the meat most commonly required.

The relative advantages of buying a chest or an upright freezer have not changed much during the last ten years. Upright freezers and fridge-freezers are easier to get at and to clean, and are, of course, economical so far as room-space is concerned. The choice of fridge-freezers, in particular, is now very wide indeed, and they can be bought in a large variety of models. Chest freezers, on the other hand, are better for accommodating large articles – a couple of casseroles, for example – and the contents suffer less if there is a power failure. Chest freezers are also slightly cheaper to run – not least because the cold air does not rush out every time the door is opened. They do, however, involve more stooping and lifting out of heavy baskets.

Small improvements have been introduced, such as internal lights, and the supply of baskets for chest freezers. One or two manufacturers provide a drainage plug fitted at the bottom of the cabinet so that when the freezer is defrosted the water drains away instead of having to be scooped out – an enormous advantage.

Which?, the consumer magazine, has published some useful articles on freezers, and it is well worth consulting these if you are thinking of replacing the one you have at present, or of buying one for the first time.

The Freezer
as Store Cupboard

****

In *The Penguin Freezer Cookbook* we gave a list of stand-bys which are useful to keep in the freezer, always ready to hand and always fresh. To this list we should like to add:

Milk in cartons. This generally freezes well, especially if you can buy the homogenized variety. It will take 6 to 8 hours to thaw at room temperature, and when it has thawed completely the carton should be given a good shake before opening, as sometimes there is a tendency for the milk to separate.

Cream. This is an extremely useful ingredient in a cook's basic store, as even a small amount stirred into a soup or sauce, casserole or egg dish will add a lot to the flavour and texture.

Commercially frozen cream, double and single, is available. This has an added stabilizer, so that it does not separate on thawing. Some brands are frozen in granules or small bars, which make it easy to add just a spoonful to a dish.

You can also freeze cream yourself. The higher the butterfat content of the cream the better it will freeze, so the best kind to buy for the freezer is thick Jersey cream or clotted cream. Any cream with less than 40 per cent of butterfat is likely to separate on thawing.

Ordinary double cream freezes well, but will keep a better texture on thawing if 1 tsp (5 ml) sugar to ½ pt (300 ml) cream is stirred in before freezing.

Whipping cream, or a combination of double and single cream, can also be frozen lightly whipped, preferably with a little sugar added.

Artistic cooks who like to decorate their food may find it useful to freeze piped rosettes of whipped cream. Open-freeze, then store in plastic boxes. These can be used straight from the freezer for decoration, as they take only 10 to 15 minutes to thaw.

As sour cream has a rather low butterfat content, it tends to separate on thawing, but can still be used for cooking. Dishes made with sour cream freeze perfectly well.

Yoghurt. This is rather tricky to freeze, as it has a tendency to separate on thawing. Some commercial yoghurts have an added stabilizer which prevents this, and the Greek 'strained' cow's or ewe's milk yoghurts also freeze well. Home-made yoghurt is less likely to separate on thawing if it has been made with full-cream milk that has been allowed to simmer for a while before the 'starter' is added. If frozen yoghurt does separate on thawing it can sometimes be reconstituted by whipping lightly, and in any case will still be suitable for cooking. Cooked dishes made with yoghurt freeze well.

Lightly whipped yoghurt can be substituted for whipped cream to make light and deliciously refreshing fruit ice creams.

Cheese. Most cheeses freeze excellently, but it is important to wrap them well, first in clingfilm and then in foil. Hard cheeses keep well for about 2 months; blue cheeses for 1 to 2 months; and cream cheeses – which should be ripe but not over-ripe when they are frozen – up to 6 months. Cottage cheese should be frozen in small quantities and mixed with a little cream after it is taken out of the freezer.

All cheese should be thawed slowly.

Grated cheese is particularly useful for freezing as it keeps almost indefinitely, and can be used straight from the freezer.

Spices. These soon lose their freshness and flavour, but keep well in the freezer. Wrap and seal them tightly in small plastic bags, making sure that each is clearly labelled, and put in one polythene bag before freezing. They can be used straight from the freezer.

Stocks

✳✳✳✳✳✳✳✳✳✳✳✳✳✳✳✳✳✳✳

Stock is to the home cook what his *fonds de cuisine* are to the professional chef: they are like a personal signature. While stock cubes are all very well, and provide a short cut which none should despise, the simplest soups and sauces made with a good home-made stock are instantly recognizable as something more special.

This is one of the areas where the freezer comes into its own as an aid to really good cooking. Stock can be stored in the freezer for an almost indefinite period of time, so that an occasional spurt of stock-making in quantity will pay off for months to come in hearty or delicately flavoured soups, in sauces made with a good reduction of stock or casseroles enriched with it; even rice, pasta and some vegetables can sometimes be transformed by being cooked in stock instead of water. You can also store bones and carcasses in the freezer until you are ready to use them; so every time you roast a chicken or other fowl, every time you bone a leg, shoulder or breast of lamb, or any time you have a good beef or veal marrow bone, and don't feel like making the stock there and then, keep the bones in a polythene bag in the freezer for future use. To extract the maximum flavour, saw large bones into pieces and crush poultry carcasses before boiling them for stock.

There are no hard and fast recipes for stock. It can be made more or less rich according to taste, inclination and availability, and the recipes that follow are only suggestions, to be adapted to individual needs. But there are a few ground rules to be observed.

Although you can make a general all-purpose stock, which will provide the base for some good hearty soups, from whatever bones or carcasses you happen to have, for most 'recipe' soups it is best to stick to either one kind of meat or a few combinations, such as chicken with any other poultry or with veal, or veal with beef, or beef with venison. Lamb or game bones are best boiled up on their own, perhaps with the addition of a chicken carcass or a calf's foot to strengthen the broth. Ham bones are generally too salty to be made into useful stock, apart from the classic split pea soup, and even then they usually need long soaking before they can safely be used.

To make stock, use a really large saucepan, cover the bones with cold water and bring slowly to the boil. A surface scum will form which should be skimmed off until no more appears. Do not add the other ingredients, such as vegetables and herbs, until you have finished skimming, otherwise you will be constantly skimming vegetables and herbs off with the scum. This procedure should be followed whether the bones have been previously roasted (see p. 6) or not.

Onions should always be added for flavouring, and carrots too if available. Other vegetables, such as tomatoes, leeks and celery, may be added according to taste, but don't add any members of the cabbage family, which will turn the stock sour. The same is true of potatoes, which should only be added at the soup stage. Garlic should not be added to the basic stock, especially if it is to be frozen, as it will give a mustiness to the flavour; it can always be added to the soup or sauce when the time comes.

Most common herbs (perhaps with the exception of sage) can be used, according to taste, but go easy with them for delicate stocks, such as chicken or veal. The best are parsley, marjoram, bay leaves or thyme. Remember that dried herbs are stronger than fresh ones, and take care with rosemary, which is particularly pungent. It is best to use only the stalks of parsley, as the stock can otherwise become rather cloudy. Never add salt to stock, as the long boiling and reducing can easily make the final result too salty.

Once the stock has been skimmed and vegetables and herbs added, allow it to simmer, uncovered, for at least an hour, but all day if you wish, provided the volume of water is sufficient. Stock can also be left to cook in the lower oven of a solid fuel cooker. When the stock is ready, always bring it to a good rolling boil for about 5 minutes before removing from the heat and straining.

Most stock needs at least two strainings, as there are usually some deposits at the bottom of the bowl after the first straining.

Before freezing, it is often worth boiling the strained stock again briskly for a while to reduce, so as to take up less freezer space.

Always allow stock to cool, then refrigerate and lift off any fat. It is best to store stock in the freezer in small quantities – ¼-, ½- and 1-pt (150-, 300- and 500-ml) tubs saved from cream or yoghurt are ideal for this. Label carefully for easy identification.

It is sometimes a good idea to reduce a really good stock until it is very concentrated indeed, to use as flavouring for sauces and casseroles to which you don't want to add too much liquid. Freeze this reduced stock in ice-cube trays, and store the frozen cubes in a polythene bag.

Concentrated meat juice or jelly from roasted joints or poultry, or left-over stews, casseroles or gravy, can also be frozen in this way and used to add flavour to soups or sauces.

Beef Stock

This is usually the best stock to use for hearty winter soups, and to add to most casseroles.

Beef bones tend to be very large, and you may want to ask your butcher to saw them up for you. Try to have at least one marrow bone among them.

Use onions, carrots, celery and if possible leeks with beef bones.

Put bones and roughly chopped vegetables in a roasting tin and roast them in a very hot oven (230°C, 450°F, gas 8) for at least 30 minutes, or until everything is a rich dark brown.

Pour off the fat, put the bones into a saucepan and cover with cold water. Bring to the boil and skim as described on p. 5.

Add the vegetables, and also some parsley stalks, bay leaves, thyme and a little rosemary. To give the stock more body, add some stewing steak or shin of beef, which will also make it a little gelatinous. Simmer for 2 to 3 hours, then strain and freeze as described above.

Chicken Stock

This is the lightest of all the stocks, and best for delicate vegetable soups, such as courgette, pea or asparagus.

You can use just the carcass and giblets left from roasting a chicken, or you can add more giblets or a few wing tips which can sometimes be bought separately. To make a really rich stock, add a boiling fowl, but remove it as soon as it is tender and use as a separate dish. You can always bone it as soon as it is cool enough to handle, and return the bones to the pot.

Use one onion and one carrot per carcass, some leek and celery if you wish, and half a lemon.

The best herbs to use are parsley stalks, thyme – especially lemon thyme – and tarragon.

Pound the carcass well with a meat pounder or rolling pin as this helps to bring out the flavour and goodness.

Put the meat and bones into a large saucepan, cover with cold water and bring to the boil. Follow instructions for skimming given on p. 5, and add vegetables and herbs when ready. Simmer for 2 to 3 hours, then strain and freeze as described on pp. 5–6.

Use the carcasses of other kinds of poultry to make stock in the same way. Duck bones make a particularly rich stock; turkey bones need plenty of lemon juice to counteract their sweetness.

Chicken and Veal Stock

This stock has slightly more body than simple chicken stock. It is ideal for chilled summer soups, as the addition of the veal makes the stock gelatinous, so that the soups will be thick and lightly set.

Add one or two veal knuckles or shin bones, and if possible a calf's foot, to the ingredients for chicken stock (see above), and follow the same procedure.

Fish Stock

Fish stock is excellent for poaching fish, for fish and some other soups, or as the basis of a fish velouté sauce and of other sauces to be served with fish. It is worth making a quantity after a visit to a friendly fishmonger who does his own filleting. Freeze in small amounts, as one often needs only a little of this stock.

1 medium onion
1 oz (25 g) butter
sprig of thyme
1 bay leaf
a few parsley stalks

12 oz (350 g) fish bones and heads
scant 1 pt (500 ml) dry white wine
scant 1 pt (500 ml) water or
    chicken stock

Sweat the sliced onion for a few minutes in the butter in a large saucepan with the thyme, bay leaf and parsley stalks. Add the fish bones and heads, the wine, and the water or stock. Simmer gently, uncovered, for 15 minutes – no more. Strain through a fine sieve and allow to cool before freezing.

Lamb Stock

This has a highly individual flavour, and is not really very useful for making other dishes. But lamb broth on its own is delicious, especially if well flavoured with oregano or marjoram, fresh if possible.

Using lamb bones, make as for beef stock, and add half a lemon and plenty of oregano or marjoram to the simmering pot.

Remove all traces of fat, and add more oregano – finely chopped if you are using the herb fresh – a little lemon juice and some finely chopped parsley before serving.

Veal Stock

Calf's feet are difficult to buy nowadays, but they are an important ingredient to give this stock body. So buy them when you can, and either keep them in the freezer until you are ready to use them, or make the stock in a good quantity and freeze it. Veal stock is useful for soups, sauces and consommés with a delicate flavour, and in particular for vichyssoise.

2 lb (1 kg) breast of veal
1–2 calf's feet
1 lb (500 g) veal bones
1–2 tbls (15–30 ml) oil
1 lb (500 g) onions
8 oz (225 g) carrots
parsley stalks
small bunch of fresh thyme or 1 tsp (5 ml) dried thyme

3 cloves garlic, unpeeled
3 tbls (45 ml) concentrated tomato purée (optional)
1¾ pts (1 L) red wine
4½ pts (2.5 L) water
12 black peppercorns

Put the breast of veal, calf's feet and bones into a roasting tin. Roast in a very hot oven (230°C, 450°F, gas 8) until golden-brown, taking care that the bones don't burn. Drain off any fat.

Heat the oil in a large saucepan, add the chopped onions and carrots, the parsley stalks, thyme and tomato purée, and cook gently until everything is well browned. Add the veal, calf's feet and bones. Pour in the wine and boil to reduce by about half. Add the water and the peppercorns and bring to a simmer.

Cook gently, uncovered, for 6 hours, stirring occasionally. Strain through a fine sieve and allow to cool. Remove the fat before freezing.

Vegetable Stock

Keep some of this stock in the freezer for making all kinds of vegetable soups – it is delicious, for instance, in mushroom soup – and vegetarian dishes. Exact quantities of the various vegetables are not important, so long as the proportions are reasonably well balanced. The following make a good combination:

roughly equal quantities of carrots, onions and the green tops of leeks

1 head celery

about 8 oz (225 g) mushrooms or mushroom stalks, or a few dried mushrooms

2 oz (50 g) butter

parsley stalks

black peppercorns

Slice the vegetables coarsely, being careful to see that every bit of grit has been rinsed out of the leeks. Melt the butter in a large saucepan and add the vegetables. Cook over a low heat for about 10 minutes, turning and stirring frequently so that they do not burn. Add the parsley stalks, a few peppercorns and enough cold water to cover.

Half-cover the pan and simmer gently for about 2 hours. Strain and allow to cool before freezing. If there is any fat, skim it off when the stock is cold.

Soups

✳✳✳✳✳✳✳✳✳✳✳✳✳✳✳✳✳✳

Soup is one of the best stand-bys of all to have in the freezer. In this section we have indicated which recipes will freeze well. The soups can be block-frozen in plastic containers and then turned out into polythene bags so that they take up less room and use fewer precious containers.

Frozen stock can be used even when the soup is to be frozen.

Chilled Almond Soup

Serves 4

This creamy white soup, of Middle Eastern ancestry, needs no cooking, although stock from the freezer should be brought to the boil. The soup is very rich, and servings should not be too large.

1 pt (600 ml) chicken or chicken and veal stock	1 tbls (15 ml) olive oil
4 oz (100 g) blanched almonds	salt and white pepper
2 cloves garlic	*1 tbls (15 ml) lemon juice*
	1 tbls (15 ml) finely chopped parsley

Bring frozen stock to the boil, boil for 1 minute, remove from the heat and leave to cool.

Place the almonds, garlic and olive oil in a blender or food processor and blend to a smooth paste. Add the stock and blend again.

Season to taste, chill in the refrigerator and add the lemon juice just before serving sprinkled with parsley.

Chilled Mushroom Soup

Serves 5–6

This rich and delicious soup can be eaten hot, but its creamy taste and texture are at their best when it is served cold.

1¾ pts (1 L) chicken stock
8 oz (225 g) mushrooms
1 tsp (5 ml) arrowroot or cornflour
3 egg yolks

¼ pt (150 ml) double or whipping
   cream
salt and freshly ground pepper

small sprigs of parsley

Thaw frozen stock over a gentle heat. Reserve a little of the stock and simmer the roughly chopped mushrooms in the remainder for about 10 minutes, in a covered saucepan, until soft. Place in a blender or food processor and blend until smooth, then return to the pan. Mix the arrowroot or cornflour with the reserved chicken stock and stir into the soup.

Beat together the egg yolks and cream in a large bowl and slowly stir in the soup. Return to the pan, season, and heat gently until the soup thickens, but do not allow it to boil.

To serve immediately: To serve hot: garnish each serving with the parsley.

To serve cold: allow to cool, then chill in the refrigerator for 3 to 4 hours before serving. Garnish as above.

To freeze: allow to cool, then freeze in a plastic container.

To serve after freezing: To serve hot: tip the frozen soup into a saucepan and heat through gently. Garnish as above. *To serve cold:* this soup takes a long time to thaw, so leave at room temperature for several hours and beat well before serving. If it is too thick, thin with a little milk or single cream. Garnish as above.

Chilled Pea Soup

Serves 6–8

Either fresh or frozen peas may be used for this refreshing summer soup.

3 pts (1.7 L) chicken or veal stock
1½ lb (750 g) peas, shelled weight if fresh
1 medium onion
2 cloves

1 clove garlic

½ pt (300 ml) double or whipping cream
salt and freshly ground pepper
finely chopped mint

Bring frozen stock gently to the boil. Add the peas, the onion stuck with the cloves, and the garlic and simmer over a low heat until the peas are tender. Remove the onion, cloves and garlic, pour into a blender or food processor and blend until smooth.

To serve immediately: chill in the refrigerator, then stir in the cream and season to taste before serving. Sprinkle each serving with a little mint.

To freeze: cool, then freeze in a plastic container.

To serve after freezing: allow to thaw, then proceed as above.

CUCUMBER SOUPS

Cucumbers have their moment of glut, like any other vegetable, and it is sometimes irritating, when the market or greenhouse is overflowing with them, that they cannot be stored in the freezer, because of their high water content. However, it is possible to make several delicious soups from cucumber which freeze very well indeed.

Chilled Cucumber and Sour Cream Soup

Serves 5–6

¼ pt (150 ml) chicken stock
1 large cucumber
¼ pt (150 ml) yoghurt
¼ pt (150 ml) sour cream
½ garlic clove

6 mint leaves
salt and freshly ground pepper
pinch of sugar
a few drops of green food colouring (optional)

Bring frozen stock to the boil, boil for 1 minute, remove from the heat and leave to cool. Cut up the unpeeled cucumber and place in a blender or food processor with the chicken stock, yoghurt, sour cream, chopped garlic and mint. Blend until smooth. Season with salt, pepper and sugar, and add a few drops of green colouring if you like, though the cucumber skin and the mint give the soup an attractive pale green colour.

Chill in the refrigerator for 2 to 3 hours before serving. If the soup is too thick, thin with a little chicken stock.

Note: this soup freezes successfully, but it is best not to add the mint leaves until ready to serve.

Thaw overnight in the refrigerator and beat well before serving. Because of the high water content of the cucumber, the soup takes a long time to thaw, so if there are still some particles of ice when you are ready to serve, place it in a saucepan over a very low heat until the particles have melted but the soup is still quite cold.

Chilled Cooked Cucumber Soup

Serves 4–6

Cooking the cucumber gives this soup a quite different flavour from the preceding recipe.

1¼ pts (750 ml) chicken or veal stock	3–4 sprigs of parsley
1 medium onion	salt and freshly ground pepper
1½ oz (40 g) butter	¼ tsp (1.2 ml) made English mustard
1 large cucumber	
6 oz (175 g) potato	*½ pt (300 ml) double or whipping cream*
	finely chopped chives

Thaw frozen stock over gentle heat. Sweat the finely chopped onion in the butter in a large saucepan until transparent. Add the cucumber, diced but unpeeled, the peeled and diced potato, stock, parsley, salt, pepper and mustard. Cover the pan and simmer for about 15 minutes until the potato is soft. Pour into a blender or food processor and blend until smooth. Chill in the refrigerator for 3 to 4 hours.

To serve immediately: stir in the cream and check the seasoning just before serving. Sprinkle each serving with a few chives.

To freeze: allow to cool, then freeze in a plastic container.

To serve after freezing: thaw in the refrigerator overnight and beat well before adding the cream. If you have to serve the soup in a hurry, you can thaw it over a very gentle heat. Garnish as above.

Lebanese Cucumber Soup

Serves 4–6

An unusual and quickly made summer soup.

¾ pt (450 ml) chicken, veal or fish stock
1 cucumber
salt
¼ pt (150 ml) tomato juice
½ pt (300 ml) yoghurt
¼ pt (150 ml) single cream

2 oz (50 g) peeled prawns, fresh or frozen (optional)
freshly ground pepper

1 clove garlic
1 tbls (15 ml) chopped mint
1 hard-boiled egg

Bring frozen stock to the boil, boil for 1 minute and leave to cool.

Peel the cucumber and cut into small dice. Salt lightly and leave in a colander for 30 minutes to drain.

Mix together the stock, tomato juice and yoghurt. (This can be done in a blender or food processor.) When quite smooth add the cucumber and cream, and the coarsely chopped prawns if you are using them. Check the seasoning. Chill in the refrigerator for 2 to 3 hours.

Cut the garlic clove in half and rub a soup tureen or individual bowls with the cut sides. Pour in the soup and just before serving sprinkle with the mint and finely chopped egg.

Avgolemono Soup

Serves 6

This light, refreshing soup, with its very lemony flavour, is equally good made with chicken, chicken and veal, or fish stock, although, strictly speaking, chicken stock is the most orthodox.

2 pts (1.2 L) stock (see above)
3 eggs
¼ pt (150 ml) lemon juice
(juice of 2–3 lemons)

1 tbls (15 ml) cooked rice
salt and freshly ground pepper

1 tbls (15 ml) finely chopped parsley or chervil

Bring the stock just to the boil.

Whisk the eggs with the lemon juice, add a little of the hot stock and whisk again. Pour the mixture slowly into the simmering stock, whisking all the time. Do not allow to boil.

Stir in the rice. Season to taste, sprinkle with the herbs and serve very hot.

Note: if you are using cooked rice from the freezer (see p. 67), add it before the eggs. Bring the soup to the boil again and boil rapidly with the rice for 2 minutes.

Chicken Curry Soup

Serves 6

A delicate soup which is equally good hot or cold – to warm you on a chilly winter's day or as a refreshing starter in the middle of a heatwave.

2 pts (1.2 L) chicken stock
1 oz (25 g) butter
1½–2 tsp (10 ml) mild curry powder
1 oz (25 g) plain flour
2 egg yolks
squeeze of lemon juice

¼ pt (150 ml) single or whipping
 cream
8 oz (225 g) cooked chicken
 (white meat)

chopped chives or parsley

Thaw frozen stock over a gentle heat.

Melt the butter in a saucepan and stir in the curry powder. Add the flour and stir over a gentle heat for 3 to 4 minutes. Slowly pour in the stock, stirring until the mixture is quite smooth, and bring to the boil. Remove from the heat and allow to cool slightly.

Beat together the egg yolks, lemon juice and cream, and stir in a little of the stock. Return to the pan and stir over a low heat until the soup has thickened slightly, but do not allow to boil.

To serve hot: add the chicken meat, cut into slivers, and heat through without allowing to boil. Sprinkle each serving with a few herbs.

To serve cold: chill in the refrigerator for several hours, and stir in the chicken meat just before serving. Garnish as above.

Stracciatella

Serves 5–6

This Italian soup gets its name – it means 'little rags' – because the beaten eggs added at the last moment break up into flakes.

2 pts (1.2 L) chicken or
  veal stock
2 eggs
2 tbls (30 ml) grated parmesan
  cheese
2 tbls (30 ml) finely grated fresh
  breadcrumbs, white or brown

2 tbls (30 ml) melted butter
freshly grated nutmeg
salt and freshly ground pepper
2 tbls (30 ml) chopped parsley

Thaw frozen stock over a gentle heat. Mix together the beaten eggs, cheese, breadcrumbs, butter and a few pinches of nutmeg. Season, then stir in about ½ cup stock. Bring the remaining stock to the boil and pour in the egg mixture all at once, whisking vigorously with a fork. Add the parsley.

Leave the pan over a moderate heat, whisking all the time, until the soup just comes to the boil. The moment it begins to show signs of bubbling pour into well-heated bowls. Serve at once.

Zuppa alla Pavese
(Soup with Egg)

This is more a light supper dish than a soup pure and simple. It is better not to try to make it for more than four, as the eggs must be watched carefully so that they are neither over- nor under-cooked. The stock needs to be well seasoned, to bring out a good flavour.

For each serving:

⅓ pt (200 ml) chicken or
  veal stock
1 egg

1 slice bread, white or brown
butter for frying
1 tbls (15 ml) grated cheese

Thaw frozen stock over a gentle heat. Put the soup plates to warm – shallow plates rather than bowls are best.

Pour the stock into a large frying pan and set over a moderate heat. When it begins to simmer, carefully break the eggs into it and poach lightly. While the eggs are cooking, gently fry the bread on both sides in a little butter until golden-brown. Cut each slice into 4 triangles.

As soon as the eggs are ready transfer one to each of the warmed soup plates and pour the stock over them. Arrange the fried bread triangles round the eggs and sprinkle with the cheese. Serve at once.

Tuscan Tomato Soup

Serves 6–8

This soup comes from the region round Florence, where the bread has a very special flavour and texture and the dark green olive oil is the best in Italy. It is quite different from the usual run of tomato soups. Outside Tuscany wholemeal bread must be used (as white bread would go mushy), and a heavy, good-quality olive oil, preferably Italian.

8 oz (225 g) wholemeal bread
1½ lb (750 g) large, ripe tomatoes
2 cloves garlic
salt
2½ pts (1.5 L) vegetable or
   chicken stock

5 tbls (75 ml) chopped parsley, or
   mixed parsley and basil
freshly ground pepper

olive oil

Cut the crusts off the bread, break it into chunks and soak in water for 30 minutes. Drain and squeeze very thoroughly in a clean tea-towel.

While the bread is soaking skin the tomatoes, halve, remove the seeds and leave cut side down to drain. Crush the garlic with a little salt. Mash the tomatoes with a fork.

Bring the stock to the boil and stir in the bread, tomato pulp, garlic and herbs. Season, cover the pan and simmer for 10 to 15 minutes.

Pour into warmed soup plates, and hand round the oil so that everyone can help themselves to a spoonful or two and mix it into the soup.

Sorrel Soup

Serves 6–8

Sorrel, with its long red stalks and pointed green acid leaves, can rarely be bought from greengrocers in this country, but it is very easy to grow. It makes an excellent sauce for fish, as well as this lovely light and refreshing summer soup, which may be eaten hot or cold. If you prefer a rather blander soup, use a mixture of sorrel and spinach.

8 oz (225 g) sorrel leaves or a mixture of sorrel and spinach
2½ pts (1.5 L) chicken stock (see p. 6)
2 oz (50 g) butter
1 large onion

1 potato
salt and freshly ground pepper

2 egg yolks
¼ pt (150 ml) cream

Wash the sorrel or sorrel and spinach leaves in several changes of cold water. Thaw frozen stock over a gentle heat.

Melt 1½ oz (40 g) of the butter in a large, heavy saucepan and cook the chopped onion and potato over a gentle heat until soft. Add the sorrel leaves, and stir over a moderate heat for a few minutes until the sorrel has wilted. Add the stock, bring to the boil and simmer for 5 minutes. Season to taste.

To serve immediately: stir the cream into the egg yolks, add a little of the soup and stir until smooth.

Add to the soup, stir over a moderate heat but do not allow to boil. Check the seasoning and, if serving hot, add the remaining butter, a little at a time, to give the soup a glaze. Omit the butter if you are serving the soup cold.

To freeze: cool, then freeze in a plastic container.

To serve after freezing: heat the frozen soup gently and then proceed as above.

Spring Vegetable Soup

Serves 6–8

It does not matter if you have rather more of one vegetable and less of another than is given in the list of ingredients, so long as they are all green and no single vegetable predominates. The soup should be very thick, but you can thin it down with extra chicken stock if you prefer.

1 lb (500 g) leeks, trimmed weight
4 tbls (60 ml) olive oil
4 oz (100 g) spinach, trimmed weight
½ small lettuce
8 oz (225 g) shelled fresh peas (or frozen)

8 oz (225 g) French beans
2 pts (1.2 L) chicken or vegetable stock
salt and freshly ground pepper
2 tbls (30 ml) chopped parsley

Wash the leeks well and chop coarsely. Heat the oil in a large saucepan and sweat the leeks for a few minutes. Add the spinach and lettuce, roughly chopped, the peas, and the French beans cut into short lengths. Cook gently for a further 5 minutes, stirring from time to time.

Meanwhile bring the stock to the boil. Add to the vegetables, season and cover the pan. Simmer for 15 to 20 minutes until the vegetables are soft.

Rub through a coarse mouli, or blend very briefly in a blender or food processor. The soup should not be puréed, as each vegetable should keep its own identity. Return to the pan to warm. Check the seasoning and stir in the parsley before serving.

Quick Tomato Soup

Serves 4

A quickly made soup for unexpected visitors. Exact quantities are not important, but the soup should have a nice spicy taste.

1 pt (600 ml) chicken or veal stock
14-oz (400-g) can tomatoes
1–2 tsp (5–10 ml) dried mixed herbs
salt and freshly ground pepper

a little sugar
about ¼ pt (150 ml) whipping cream

chopped parsley

Thaw frozen stock over a gentle heat.

Place the tomatoes in a saucepan with the stock and herbs and season with salt, pepper and a little sugar. Bring to the boil and boil for 5 minutes. Liquidize in a blender or food processor, then strain. Return to the pan and stir in the cream. Check the seasoning and serve sprinkled with parsley.

Potato Soup

Serves 8

3 pts (1.7 L) chicken, veal or
vegetable stock
4 medium leeks or 1 lb (500 g) onions
or a mixture of the two
2 oz (50 g) butter
2 lb (1 kg) potatoes
salt and freshly ground pepper

5–6 tbls (75–90 ml) cream
2 oz (50 g) butter
milk (optional)
chopped parsley
croûtons (optional)

Thaw frozen stock over a gentle heat. Slice the white part of the leeks, or the onions, and cook lightly in the butter. Add the potatoes, peeled and roughly cut up, and the stock. Season, cover and simmer gently for about 15 minutes until all the vegetables are soft. Place in a blender or food processor and blend until smooth.

To serve immediately: return to the pan and stir in the cream and the butter, cut into small pieces. If the soup is too thick, thin with a little milk. Check the seasoning. Sprinkle each serving with a little parsley, and serve very hot, with croûtons if you like.

To freeze: allow to cool, then freeze in a plastic container.

To serve after freezing: tip the frozen soup into a saucepan and heat through very gently. Proceed as above.

Cream of Lentil Soup

Serves 6–8

8 oz (225 g) lentils
1½ pts (900 ml) chicken stock
2 medium onions
1 large carrot

salt and freshly ground pepper

3–4 tbls (45–60 ml) cream
chopped parsley

Soak the lentils in cold water for about 1 hour. Meanwhile thaw frozen stock over a gentle heat.

Drain the lentils. Put them into a saucepan with the chopped onions, carrot and stock. Season and bring to the boil, then simmer gently, covered, for 30 to 45 minutes, until the lentils are puréed and the vegetables are soft. Pour into a blender or food processor and blend until smooth.

Return to the saucepan and reheat, adding more chicken stock if necessary – but the soup should be quite thick. Check the seasoning.

To serve immediately: add the cream and sprinkle a little parsley over each serving.

To freeze: allow to cool, then freeze in a plastic container.

To serve after freezing: tip the frozen soup into a saucepan and heat very gently, stirring from time to time. Garnish as above.

Note: a few bacon rinds added to the pan with the vegetables add extra flavour. The soup is also good served with croûtons of bread rubbed with garlic and fried until golden.

Goulasch Soup

Serves 6–8

A thick, warming soup, almost a meal in itself. The quantity of paprika given here is approximate, as it varies so much in strength.

2½ pts (1.5 L) beef stock
8 oz (225 g) shin of beef
1 tbls (15 ml) plain flour
salt and freshly ground pepper
2 tsp (10 ml) paprika
½ oz (15 g) butter
2 tsp (10 ml) oil

2 onions
1 large clove garlic
2 tsp (10 ml) tomato purée
1 oz (25 g) dried mushrooms
 (optional)
1–2 red peppers, fresh or canned

¼ pt (150 ml) sour cream

Thaw frozen stock over a gentle heat. Cut the beef into small cubes, discarding any fat or gristle. Roll the beef cubes in the flour seasoned with salt, plenty of pepper and the paprika.

Heat the butter and oil in a large saucepan and brown the beef on all sides. Add the finely chopped onions and garlic, cover and cook over a gentle heat until the onions are soft. Stir in any remaining seasoned flour, add the tomato purée and blend well. Slowly stir in the stock, add the mushrooms, if you are using them, cover and simmer for about 2 hours, or until the beef is tender. Check the seasoning.

If you are using fresh red peppers, core, seed and blanch them for 2 minutes. Cut the peppers into thin strips and add to the soup.

To serve immediately: stir in the sour cream and serve very hot.

To freeze: allow to cool, then freeze in a plastic container.

To serve after freezing: heat the soup gently from frozen. Check the seasoning (you will probably need to add a little more paprika at this point). Proceed as above.

Cream of Fish Soup

Serves 6

2 pts (1.2 L) fish or chicken stock
¼ pt (150 ml) dry white wine
1 lb (500 g) fillets of plaice, haddock, whiting or cod
1½ oz (40 g) butter

1½ oz (40 g) plain flour
salt and freshly ground pepper

¼ pt (150 ml) double or whipping cream
chopped parsley

Bring the stock to the boil, add the wine and fish, and simmer gently for a few minutes until the fish is cooked. Strain, reserving the stock, and remove any skin from the fish.

Melt the butter in a saucepan, add the flour, and stir over a low heat for 2–3 minutes. Gradually add the reserved stock, stirring until the sauce is thickened and smooth. Simmer for a few minutes, then blend with the fish until creamy. Season.

To serve immediately: return to the pan, stir in the cream and bring to the boil. Sprinkle with parsley before serving.

To freeze: allow to cool, then freeze in a plastic container.

To serve after freezing: tip the frozen soup into a saucepan and bring gently to the boil. Proceed as above.

Sole Velouté Soup

Serves 6

This is a delectable soup, and though it sounds extravagant it is quite filling, so a little goes a long way.

1 lb (500 g) sole fillets, fresh or frozen
4 oz (100 g) mushrooms
3 oz (75 g) butter
1 tbls (15 ml) cornflour
½ pt (300 ml) double or whipping cream

3 egg yolks
1 pt (600 ml) fish stock
salt and white pepper

chopped parsley

If using frozen fish, allow to thaw. Skin the sole fillets and cut them into small lozenges about 2 × ½ in (5 × 1 cm).

Sweat the thinly sliced mushrooms in 2 oz (50 g) of the butter. Set aside.

Mix the cornflour with a spoonful or two of the cream, then add the beaten egg yolks and the remaining cream. Bring the fish stock to the boil in a large saucepan and slowly stir a little of it into the egg and cream mixture. Return to the pan and reheat, stirring. Be careful not to allow the mixture to boil. It should have the consistency of cream, so add a little more fish stock, cream or milk if it is too thick. Season lightly. Stir in the remaining butter, cut into small pieces, and the mushrooms.

In a separate pan poach the sole lozenges in just enough water to cover. They will be ready almost as soon as the water has come to the boil. Drain well and add to the velouté.

Serve in individual heated soup bowls, sprinkled with a little parsley.

Cheese and Onion Soup

Serves 6

A really strong, sustaining soup. Ideally, it should be made with gruyère cheese, but a good-quality cheddar will do.

1¾ pts (1 L) beef stock	6 thick slices stale French bread
4 oz (100 g) butter	6 oz (175 g) gruyère or
1 lb (500 g) onions	cheddar cheese
1 oz (25 g) plain flour	2 egg yolks
pinch of dried mixed herbs	1–2 tbls (15–30 ml) brandy
salt and freshly ground pepper	

Thaw frozen stock over a gentle heat.

Peel and finely chop the onions.

Melt the butter in a large, heavy saucepan and add the onions. Cover and cook over a moderate heat until soft and golden. Do not allow to brown.

Sprinkle on the flour, stir well and cook, stirring, for a further 2 minutes. Slowly stir in the stock. Add the herbs, salt and pepper. Bring to the boil, cover and simmer for at least 30 minutes. Pour the soup into a blender or food processor, blend until smooth and check the seasoning.

While the soup is cooking, heat the oven to 220°C, 425°F, gas 7. Dry out the pieces of bread in the oven or toast them quite lightly.

Place the bread in an ovenproof tureen or casserole, sprinkle on half the cheese and pour on the soup. Sprinkle the remaining cheese on top. Place the tureen near the top of the oven and heat for 10 to 15 minutes or until the cheese has melted and begins to form a thick crust.

Whisk the egg yolks with the brandy and pour into the soup. Give a good stir and serve very hot.

Stilton Soup

Serves 6

A thick, velvety soup with a surprisingly mild flavour. A good way to use up stilton left over from Christmas.

¾ pt (450 ml) chicken stock
2 oz (50 g) butter
1 oz (25 g) plain flour
¾ pt (450 ml) milk

6 oz (175 g) stilton
salt and freshly ground pepper

1 tbls (15 ml) cream
1 tbls (15 ml) finely chopped parsley

Thaw frozen stock over a gentle heat.

Melt the butter in a large, heavy saucepan, add the flour, stir well to blend and cook, without allowing to brown, for 3 minutes.

Slowly add the milk, stirring to keep the mixture smooth. Add the stock and bring to the boil.

Remove from the heat and crumble in the stilton. Add seasoning to taste, return to a moderate heat and stir well. Allow to simmer for at least 5 minutes.

Just before serving stir in the cream and sprinkle with parsley. Serve very hot.

Pâtés

✳✳✳✳✳✳✳✳✳✳✳✳✳✳✳✳✳✳✳

A pâté is an excellent stand-by to have in the freezer, especially over Christmas or other holiday periods. It can be served as a lunch dish with salad, or as a first course on more formal occasions, and smooth pâtés can be used to make snacks and cocktail canapés.

If you are using frozen livers (and a pack of frozen chicken livers is an indispensable 'store cupboard' item in any freezer) they must thaw completely before cooking, as otherwise they will not cook evenly, and may be overdone on the outside before the inside has set. An 8-oz (225-g) tub takes approximately 3 hours to thaw at room temperature, or overnight in the refrigerator. Leave in the closed container to thaw, and, once thawed, use as soon as possible. Pick the livers over before using, in order to discard any membranes or discoloured or greenish parts.

Pâté made with frozen meat may be frozen again, as the meat will have been cooked in the meantime, but it is best not to keep pâté in the freezer for too long (a few weeks at the most). However, fish pâtés made with frozen fish should be eaten fresh and not refrozen.

Pâté should be frozen in pots or dishes of a size that you are likely to use in one or at most two sittings. Thaw pâtés in the refrigerator – 24 hours should be ample for large ones, 1 to 2 hours for a ramekin-sized one. Eat within 24 hours.

Chicken Liver Pâté

A very rich, smooth pâté, which can be further enriched by the addition of duck or any other poultry liver. It is particularly good to make before Christmas if you have a goose liver, which makes the richest pâté of all.

1 lb (500 g) poultry livers
8 oz (225 g) butter
1 clove garlic
2 tbls (30 ml) brandy
2 tbls (30 ml) madeira or sherry

good pinch of dried mixed herbs
salt and freshly ground pepper
bay leaves

Allow frozen livers to thaw, and prepare as described on p. 25.

Melt half the butter in a heavy frying pan, add the livers with the roughly chopped garlic and cook over a moderate heat for 5 minutes, stirring so that the livers cook evenly. They should be brown on the outside, and still pink but set in the centre.

Pour the contents of the frying pan into a blender or food processor. Add the brandy and madeira or sherry to the pan and bring quickly to the boil, stirring well to loosen and amalgamate any sediment left on the bottom of the pan. Allow to bubble for 1 minute, then add to the livers. Add the herbs and blend until smooth. Blend in the remaining butter and salt and pepper to taste. Be careful not to overblend if you are using a food processor.

Pour into one or several earthenware terrines or ramekin dishes and place a bay leaf on top.

To serve immediately: refrigerate overnight before serving to allow the pâté to set and the flavour to mature. If the pâté is to be kept in the refrigerator for several days before serving, pour enough clarified, melted butter over the top to form a complete seal. The pâté will then keep for at least a week.

To freeze: seal the pâté with clingfilm, wrap the terrine or ramekin dishes in foil, and freeze.

To serve after freezing: allow to thaw for several hours in the refrigerator. Serve within 24 hours of thawing.

Chicken Liver Pâté with Gin

1 lb (500 g) poultry livers
2 tbls (30 ml) butter
2 cloves garlic
4 tbls (60 ml) gin

8–10 juniper berries
¼ pt (150 ml) double cream
salt and freshly ground pepper

Allow frozen livers to thaw, and prepare as described on p. 25.

Sauté the livers in the butter until they are just firm, adding the chopped garlic to the pan near the end of the cooking time. Put into a blender or food processor. Deglaze the pan with the gin, and add the liquor from the pan to the livers, together with the crushed juniper berries and the cream. Blend until smooth. Season and pour into one or several earthenware dishes, terrines or ramekin dishes.

To serve immediately: leave in the refrigerator for a few days if possible, as the flavour matures and improves with keeping.

To freeze: seal the pâté with clingfilm, wrap the terrine or ramekin dishes in foil, and freeze.

To serve after freezing: allow to thaw for several hours in the refrigerator. Serve within 24 hours of thawing.

Hot Chicken Liver Pâté

This pâté is good either with drinks or as a starter.

8 oz (225 g) chicken livers	3–4 anchovies
1 tbls (15 ml) olive oil	2 tsp (10 ml) capers
1 oz (25 g) onion	1 egg yolk
1 medium carrot	1–2 tbls (15–30 ml) dry white wine
1 celery stalk	knob of butter
1 clove garlic	salt and freshly ground pepper
few sprigs of parsley	a little stock (optional)
grated rind of 1 lemon	

Allow frozen livers to thaw, and prepare as described on p. 25.

Heat the oil in a small saucepan and gently fry the finely chopped onion, carrot and celery, crushed garlic, parsley and lemon rind until the onions are soft and golden. Add the chopped chicken livers, cover and cook gently for 10 minutes.

Pour into a blender or food processor together with the anchovies, capers, egg yolk, wine and butter and blend until smooth. Season. Return to the pan and heat gently. If the mixture seems too stiff, add a little stock.

To serve immediately: serve spread on thin slices of hot toast, cut into fingers or triangles. The pâté should be served as hot as possible.

To freeze: allow to cool, then freeze in a plastic container.

To serve after freezing: warm the frozen pâté over a very low heat and serve as above.

Chicken and Lemon Terrine

This is a light and refreshing summer terrine, especially good for a buffet party. Since it is a little complicated to prepare, it is a good idea to make it in advance and keep it in the freezer. However, as the flavour is delicate, do not keep the terrine in the freezer for more than a week or two.

Serves 12–14

1 large roasting chicken
    [about 4½ lb (2 kg)]
8 oz (225 g) belly of pork
8 oz (225 g) lean veal
2 eggs
2–3 tbls (30–45 ml) brandy
salt and freshly ground pepper
pinch of freshly grated nutmeg

pinch of ground cloves
knob of butter
1 lemon
2 oz (50 g) fresh parsley
8 oz (225 g) streaky bacon

sprigs of parsley
slices of lemon

Skin the chicken – once you have inserted your fingers under the skin along the ridge of the breastbone you will find that you can pull the skin off easily. With a sharp knife cut off the breast meat into slices a good ⅛ in (2.5 mm) thick, and set aside. Do the same as far as you can with the drumsticks and thighs, discarding any bits of inner skin or tendons, and set these aside also. Remove as much as possible of the remaining meat from the carcass.

Remove the skin and any bits of bone from the belly of pork. (These will make an excellent stock, together with the chicken skin and carcass and the bacon rinds.)

Mince the small pieces of chicken meat, the chicken liver, heart and gizzard together with the pork and veal.

Put the minced meat into a large bowl, add the lightly beaten eggs, brandy, salt, pepper and spices and mix well. Fry a small pat of the mixture in a little butter to test the seasoning.

Cut the lemon in half crossways and scoop out the flesh with a grapefruit knife or teaspoon. Add to the meat mixture. Chop the parsley quite coarsely and add also.

Remove the rinds from the bacon rashers and completely line a 2-pt (1.2-L) terrine or a 2-lb (1-kg) loaf tin with the bacon slices. Spoon in half the minced meat mixture, press down into a smooth layer with the back of a spoon, lay the chicken slices on top and cover with the remaining minced meat. Top with any remaining bacon slices, and cover with a double thickness of foil.

Stand the terrine or loaf tin in a *bain-marie* and cook in a moderate oven (180°C, 350°F, gas 4) for approximately 2½ hours. The terrine is done when the juices that rise to the top are just clear and the pâté has shrunk slightly away from the sides.

Remove from the oven, allow to cool slightly and then weight down with a wooden board and heavy weights (cans will do if you have no weights).

To serve immediately: the pâté should be allowed to mature overnight in the refrigerator. Turn out on to a serving dish and decorate with sprigs of parsley and slices of lemon.

To freeze: when the pâté has cooled and set, either turn out of the terrine or tin, wrap in clingfilm and foil and freeze, or leave in the dish, overwrap and freeze.

To serve after freezing: remove from the freezer 24 hours ahead and leave in the refrigerator overnight. Allow an hour or two at room temperature before serving as above.

Duck Galantine

Serves 10–12

Like the preceding recipe, this also looks beautiful on a buffet table. As it is a little time-consuming to make and freezes well, it is an excellent dish to make ahead of time for a party.

Unless you are very skilled, it is best to get the butcher to bone the duck, and if possible also to mince the other meats for you.

1 duck, boned
1 duck liver
4 oz (100 g) chicken livers
8 oz (225 g) shallots
salt and freshly ground pepper
pinch of ground allspice
sprig of thyme
2 bay leaves

3 tbls (45 ml) calvados or brandy
1 oz (25 g) butter
8 oz (225 g) lean pork, minced
8 oz (225 g) pork fat, minced
8 oz (225 g) veal, minced
1 egg
3–4 strips pork fat or streaky bacon

Lay the boned duck breast-side down on a board and slit the skin lengthways down the middle of the back. With your fingers or a small sharp knife strip away most of the flesh, being careful not to break or pierce the skin. You will find that it comes away fairly easily, especially from the two pieces of breast. Remove the duck skin and set aside.

Cut the meat into strips and place in a bowl, together with the livers also cut into large strips, and 2 or 3 halved shallots, salt and pepper, the allspice, thyme and one bay leaf. Sprinkle on the calvados or brandy and leave to marinate overnight, or for a few hours at least, then strain, reserving the liquid.

Chop the remaining shallots and fry in the butter until soft and golden. Put the minced pork, pork fat and veal into a bowl and add the cooked shallots and the egg beaten with the strained marinating liquid. Mix well, fry a little of the mixture in some butter to test the seasoning, and season as necessary.

Flatten out the duck skin on a work surface, outside down, and spread half the forcemeat mixture down the centre. Cover with the strips of duck meat and livers, then top with the remaining forcemeat. Pat into a rounded shape, then wrap the sides of the duck skin right round. Sew up the centre, then sew up all the other openings.

Heat the oven to 180°C, 350°F, gas 4.

Season the outside of the duck parcel lightly, put it into an oval terrine just large enough to hold it, and place the second bay leaf and the strips of pork fat or bacon on top. Cook in the oven for 1½ hours, or until the juice runs just clear when the duck is pricked. Place a piece of buttered paper, butter side down, over the galantine, put a board on top, and then a heavy weight to press the galantine down.

To serve immediately: leave overnight to set and allow the flavour to mature. You can serve the galantine from the terrine, or unmould it before serving.

To freeze: freeze the galantine either in its terrine or unmoulded and wrapped in foil.

To serve after freezing: allow to thaw in the bottom of the refrigerator overnight, and leave at room temperature for 2 to 3 hours before serving.

Pork Terrine

Serves 8–12

The pig's trotter is optional, but it enriches the flavour and thickens the juices of this terrine, which freezes extremely well.

2–2¼ lb (1 kg) belly of pork, skinned and boned	1 egg
	2 tbls (30 ml) brandy
8 oz (225 g) pig's liver	salt and freshly ground pepper
1 onion	1 pig's trotter, split in half
2 cloves garlic	2 bay leaves
2 sprigs of fresh rosemary	

Heat the oven to 220°C, 425°F, gas 7.

Mince the pork coarsely with the liver. Add the finely chopped onion and garlic and the leaves of one of the rosemary sprigs to the meat mixture.

Beat the egg with the brandy and season generously. Mix into the meat and fry a small pat of the mixture to test for seasoning. Press the mixture evenly into a 2-pt (1.2-L) terrine.

Blanch the pig's trotter in boiling water for 2 minutes, then drain and press into the top of the terrine, together with the second sprig of rosemary and the bay leaves. Place the terrine in a roasting tin, pour in cold water to come halfway up the terrine and place uncovered in the oven.

Cook for 15 minutes, then lower the oven heat to 190°C, 375°F, gas 5, and continue to cook until the meat has shrunk away from the sides of the dish and the cooking juices run clear.

Remove the trotter, smooth the top of the terrine and rearrange the rosemary and bay leaves.

Leave to cool slightly, then weight down with a board and a heavy weight on top.

To serve immediately: leave in the refrigerator for at least a day to allow the jelly to set and the flavour to mature. Remove from the refrigerator 1 hour before serving. It is best to serve this pâté in its terrine.

To freeze: when the pâté has completely cooled, wrap the whole terrine in foil and freeze.

To serve after freezing: allow to thaw in the bottom of the refrigerator overnight, and leave at room temperature for 2 to 3 hours before serving.

Country Pâté

This is a robust pâté, particularly useful if you find the taste of pig's liver too strong.

1 lb (500 g) lamb's liver	¼ tsp (1.2 ml) ground cloves
1 lb (500 g) lean fresh pork	pinch of ground mace
1 lb (500 g) salt pork	salt and freshly ground pepper
1 medium onion	1 bay leaf
¼ pt (150 ml) dry white wine	

Heat the oven to 120°C, 250°F, gas ½.

Mince the liver with the pork and salt pork. Add the very finely chopped onion together with the wine, cloves and mace. Stir well together and season generously. Fry a small pat of the mixture in a little butter to test for seasoning.

Put the mixture into a 2-pt (1.2-L) terrine and place the bay leaf on top. Over this lay a piece of greaseproof paper and then the lid, or cover closely with a piece of foil. Place the terrine in a roasting tin, pour in boiling water to come halfway up the terrine and cook for 3½ hours, until the pâté has shrunk away from the sides of the terrine and the cooking juices run clear.

Leave to cool slightly, then weight down with a board and a heavy weight on top.

To serve immediately: leave in the refrigerator for at least a day to allow the jelly to set and the flavour to mature. Remove from the refrigerator 1 hour before serving, and serve in the terrine.

To freeze: when the pâté has completely cooled, wrap the whole terrine in foil and freeze.

To serve after freezing: allow to thaw in the refrigerator overnight, and leave at room temperature for 2 to 3 hours before serving.

Kipper Pâté

This is very quickly made from frozen kipper fillets. Serve with hot toast as a tasty but inexpensive starter, or spread on croûtons of fried bread or small water biscuits to serve as canapés with drinks.

2 × 8-oz (225-g) packets frozen
  kipper fillets
4 oz (100 g) butter
1 tsp (5 ml) creamed horseradish

juice of ½ lemon
1 tbls (15 ml) sherry (optional)
freshly ground pepper

Cook the frozen kipper fillets as directed on the packet, leave to cool, then skin.

Put the kippers into a blender or food processor with all the remaining ingredients. Blend until just smooth. Pour into a bowl or small individual dishes and chill in the refrigerator for an hour or two before serving.

Note: do not freeze this pâté if you have made it from frozen kipper fillets.

Smoked Mackerel Pâté

2 medium smoked mackerel
pinch of ground mace
3 oz (75 g) butter, melted
8 oz (225 g) curd cheese

a little lemon juice
freshly ground pepper
1 tsp (5 ml) creamed horseradish
  (optional)

Remove the skin and bones from the mackerel. Put the flesh into a blender or food processor with the mace, butter and curd cheese, and blend until smooth. Add lemon juice and pepper to taste, and the horseradish if you are using it.

To serve immediately: put into a bowl or terrine and chill in the refrigerator for an hour or so before serving. Serve with lemon wedges.

To freeze: freeze in foil or plastic containers, sealed very tightly so that the fishy smell does not penetrate into the rest of the freezer.

To serve after freezing: thaw in the refrigerator overnight. Serve as above.

Note: this pâté is best eaten within a month or two of freezing.

Do not freeze it if you have made it from frozen smoked mackerel.

Smoked Salmon Pâté

8 oz (225 g) smoked salmon
   trimmings
¼ pt (150 ml) double cream

6 oz (175 g) unsalted butter
lemon juice
freshly ground pepper

Put the roughly chopped smoked salmon trimmings into a blender or food processor together with the cream and the butter, which should be at room temperature. Blend until smooth. Add a little lemon juice, and season to taste with pepper. Put into one or more small terrines or bowls.

To serve immediately: serve with lemon wedges and brown bread and butter.

To freeze: cover well and freeze.

To serve after freezing: allow to thaw overnight in the refrigerator. Serve as above.

Note: do not freeze this pâté if you have made it from frozen smoked salmon.

Fish

✳✳✳✳✳✳✳✳✳✳✳✳✳✳✳✳✳✳✳✳

In the first *Penguin Freezer Cookbook* we said that fish is really only worth freezing at home if it practically jumps from sea or river into the freezer. This is still true, but during the past ten years the fishing industry has improved its freezing techniques so greatly that there is little seafood which cannot be bought frozen and eaten with enjoyment. Great care is taken to see that the fish is frozen while it is still extremely fresh, and the process is carried out in such a way that the fish comes out of its packet in prime condition. Meanwhile, fishmongers have been disappearing rapidly from our high streets, making it ever more difficult to find a good variety of 'wet' fish – though here and there they are beginning to make a welcome reappearance. While frozen fish cannot compare with the truly fresh, it is infinitely preferable to the limp and tired objects which are still all too often found on fishmongers' slabs. Moreover, the range of frozen fish is much larger than it used to be: today it is possible to buy frozen salmon, cod, haddock, plaice, lemon and dover sole, whiting and coley, as well as golden haddock cutlets, smoked mackerel fillets and kippers. An increasing range of shellfish is also becoming available, from shrimps, prawns and crabs to mussels, scallops, lobsters, crayfish and cockles.

With a selection of these in the freezer you can make a great variety of dishes, ranging from the simple to the quite luxurious and exotic. There is the added advantage that most frozen fish is best cooked straight from the freezer, before it is thawed, so that many recipes can be prepared at short notice. But one word of warning: don't keep fish in the freezer for more than 3 months, since the flavour and texture deteriorate.

No freezing instructions are given for the recipes (except for the fishcakes), as they all use frozen fish, which should not be refrozen.

Baked Fish

Serves 4

One of the quickest and simplest but none the less most delicious ways of using frozen white fish fillets. Cod, haddock, plaice or sole will do equally well.

1½–2 lb (675–900 g) frozen white
    fish fillets
2 oz (50 g) butter
salt and freshly ground pepper
1 onion or 2–3 shallots

1 lemon
¼ pt (150 ml) dry white wine

1 tbls (15 ml) chopped parsley

There is no need to thaw the frozen fish.

Heat the oven to 180°C, 350°F, gas 4.

Butter a large flameproof casserole very thickly and arrange the fish fillets in it as far as possible in one layer. Season the fish and sprinkle with the finely chopped onion or shallots. Slice the lemon and place a slice on each piece of fish.

Pour on the wine, bring gently to the boil on top of the stove, then cover and cook in the oven for about 12 minutes until the fish is cooked through.

Remove the casserole from the oven and carefully transfer the fish to a heated serving plate.

Bring the liquid in the casserole to the boil, boil briskly to reduce by half, and add the remaining butter.

Pour the sauce over the fish and sprinkle with the parsley.

Baked Fish Provençale

Serves 4

The contrast of colours makes this a very pretty dish. The pimientos and onions will be slightly crunchy, so soften them in a little butter first if you prefer.

1½–2 lb (675–900 g) frozen haddock
   or plaice fillets
1 green pimiento
1 red pimiento
1 medium onion
salt and freshly ground pepper
2 large tomatoes
1 lemon

2 tbls (30 ml) dry vermouth
6 tbls (90 ml) dry white wine
6 tbls (90 ml) fish stock
2 oz (50 g) butter, cut into small
   pieces

2 tbls (30 ml) chopped parsley
2 tbls (30 ml) chopped chives

There is no need to thaw the frozen fish.

Heat the oven to 200°C, 400°F, gas 6.

Wash and seed the pimientos and chop them finely, together with the onion.

Butter a large, shallow flameproof casserole and spread the pimientos and onion on the bottom. Lay the fish on top and season with salt and pepper. Skin, halve and seed the tomatoes and chop roughly. Blanch the lemon for 1 minute and slice finely. Lay the tomatoes and lemon on top of the fish. Pour in the vermouth, wine and stock. Cover the casserole and set over a low heat. When it begins to simmer transfer to the oven and cook until the fish is cooked through (10 to 15 minutes).

Discard the lemon slices and remove the fish to a heated serving dish. Strain the cooking liquid into a small saucepan. Arrange the strained vegetables round the fish and keep warm while you make the sauce.

Reduce the cooking liquid to about 2 tbls (30 ml) over a high heat. Remove the pan from the heat and beat in 2 pieces of butter. Return the pan to a low heat and continue to add the butter, a piece at a time, vigorously whisking with each addition, until the sauce is smooth and creamy.

Pour the sauce over the fish and serve at once, sprinkled with the parsley and chives.

Fillets of Fish with Onion Sauce

Serves 4

Another excellent way with frozen white fish fillets – cod, haddock, lemon sole or plaice.

1½–2 lb (675–900 g) frozen white
   fish fillets
¼ pt (150 ml) white wine or cider
8 oz (225 g) onions
½ lemon
1 bay leaf
a few sprigs of parsley

2½ oz (65 g) butter
¼ pt (150 ml) double cream
pinch of freshly grated nutmeg
½ tsp (2.5 ml) French mustard
salt and freshly ground pepper
2 tbls (30 ml) fine dry breadcrumbs

There is no need to thaw the frozen fish.

Put the fish fillets into a large saucepan. Add the wine or cider, 2 large slices of onion, one slice of the lemon, the bay leaf and parsley, add just enough cold water to cover, and bring very slowly to a simmer. Poach until the fish is cooked.

Remove from the heat, lift the fish out gently with a slotted spoon, drain well and place in a buttered gratin dish. Keep warm.

Bring the cooking liquid to the boil and boil rapidly for a few minutes to reduce, then strain.

Chop the remaining onions finely. Melt 2 oz (50 g) of the butter in a heavy saucepan, add the onions, cover and cook very gently until quite soft and almost puréed, but barely coloured.

Add 4 tbls (60 ml) of the strained fish stock to the onions, then stir in the cream, nutmeg, mustard, a good squeeze of lemon juice, and salt and pepper to taste. Pour the sauce over the fish, sprinkle with the breadcrumbs and dot with the remaining butter. Brown quickly under a hot grill before serving.

Freeze any remaining fish stock.

Hungarian Fish

Serves 4

Any white fish fillets can be used for this dish – cod, haddock, lemon sole or plaice.

1½–2 lb (675–900 g) frozen white
    fish fillets
1 medium onion
1 tbls (15 ml) oil
12 oz (350 g) mushrooms
2 oz (50 g) butter

1–2 tsp (5–10 ml) mild paprika
⅓ pt (200 ml) fish stock
6 tbls (90 ml) sour cream
salt and freshly ground pepper

chopped parsley

There is no need to thaw the frozen fish.

Sweat the finely chopped onion in the oil in a saucepan until soft and transparent. Add the sliced mushrooms to the pan, together with the butter. Cook gently for a few minutes. Stir in the paprika and add the fish stock. Cover and simmer for 10 minutes. Remove the pan from the heat, allow to cool for a few minutes, then gradually stir in the sour cream. Season to taste.

Butter a flameproof casserole. Lay the fish in it and pour over the sauce. Cover and leave over a very low heat until the fish is cooked (about 30 minutes). Sprinkle with parsley before serving.

Fish Pie

Serves 6–8

Fish and parsley sauce always go well together, and this easy recipe brings out the flavour of both.

2 lb (900 g) frozen cod or
    haddock fillets
3 lb (1.5 kg) potatoes
1¼ pts (750 ml) milk
4 oz (100 g) butter
2 oz (50 g) plain flour

salt and freshly ground pepper
4–6 tbls (60–90 ml) chopped parsley
    (or more)
butter and hot milk, for mashing the
    potatoes

There is no need to thaw the frozen fish.

Heat the oven to 180°C, 350°F, gas 4.

Boil the potatoes in salted water. Meanwhile gently simmer the fish in the milk for 15 to 20 minutes. When it is cooked, drain, reserving the milk, remove the skin from the fish and flake the flesh. Alternatively, place the fish in the saucepan, add the milk, bring slowly to the boil and as soon as it has boiled remove from the heat, cover the pan and leave. The fish will be cooked by the time it is cool enough to handle.

Melt 3 oz (75 g) butter in a saucepan, add the flour and stir for 2 to 3 minutes over a gentle heat. Gradually add the reserved milk, stirring all the time, until the sauce is thick and smooth. Season and stir in the chopped parsley – the more the better.

When the potatoes are cooked, drain and return to the pan. Add plenty of butter and hot milk, beating well until light and fluffy. Season generously with pepper (one of the secrets of good mashed potatoes).

Stir the fish carefully into the sauce and transfer to a heated ovenproof dish. Spoon the mashed potatoes evenly over the fish.

Dot with the remaining butter and heat through in the oven for about 30 minutes, until hot and bubbling. Put under a hot grill for a few minutes before serving so that the top is golden-brown and slightly crusty.

Norwegian Fish Pie

Serves 6–8

An excellent family dish.

2 lb (900 g) frozen white fish fillets – cod, haddock, plaice
¾ pt (450 ml) milk
about ½ pt (300 ml) water
1 onion
1 bay leaf
blade of mace
3 oz (75 g) butter
2 oz (50 g) plain flour
1 × 10-oz (300-g) can concentrated mushroom, sweetcorn or scallop soup (optional)

½ pt (300 ml) white wine or water
1 tsp (5 ml) anchovy essence
4 oz (100 g) frozen peeled prawns
4 hard-boiled eggs
salt and freshly ground pepper
3 lb (1.5 kg) potatoes, mashed
1 tbls (15 ml) dried breadcrumbs

Place the frozen fish in a wide saucepan, add the milk and just enough water to cover the fish. Add the quartered onion, the bay leaf and mace.

Bring slowly to the boil, then immediately remove from the heat, cover and leave. The fish will be cooked by the time it is cool enough to handle. Lift out with a slotted spoon, skin if necessary and flake the flesh. Place in a pie dish. Strain the cooking liquid.

Melt 2 oz (50 g) of butter in a clean saucepan, add the flour and stir for 2 to 3 minutes over a gentle heat. Blend in the strained fish liquor. Add the soup if you are using it and ½ pt (300 ml) wine or water, blend well and bring briefly to the boil. Remove from the heat.

Add the anchovy essence, prawns and hard-boiled eggs, each cut into eighths. Season to taste and pour over the fish in the pie dish. Cover with a layer of mashed potatoes (see Fish Pie, p. 40).

Heat the oven to 190°C, 375°F, gas 5. Dot the pie with the remaining butter, sprinkle with the breadcrumbs and cook in the centre of the oven for 30 to 40 minutes, until the top is a crunchy golden-brown.

Fish Poached in Vermouth

Serves 4

The dry vermouth in which the fish is cooked gives this dish a subtly different taste.

1½–2 lb (675–900 g) frozen white
    fish fillets – cod, haddock, plaice or
    lemon sole
½ pt (300 ml) dry vermouth
3 egg yolks

3 oz (75 g) butter
2–3 tbls (30–45 ml) cream
salt and freshly ground pepper

sprigs of parsley

Arrange the frozen fish, as nearly as possible in one layer, in a large saucepan. Pour over the vermouth. Cover and poach very gently for about 20 minutes until the fish is cooked through. Transfer the fish to a shallow flameproof dish and keep warm (but do not allow the fish to become dry).

Strain the cooking liquor, return to the pan and boil rapidly to reduce to 2 tbls (30 ml).

Put the egg yolks and the butter, cut into small pieces, in a basin or the top of a double boiler over hot (but not boiling) water. Stir well until the butter has melted and the mixture has thickened. (You can also make this sauce in a blender or food processor. Blend the yolks briefly. Heat the butter until it just begins to froth and, with blender at top speed, pour slowly on the yolks as you would for making mayonnaise.) Stir in the cream and the reduced cooking liquor. Season. When the mixture is creamy, use to coat the fish, and put the dish under the grill for a minute or two to brown.

Serve at once, garnished with parsley.

Fish Cakes

Makes about 16 fish cakes

These fish cakes can be made with fresh fish for freezing uncooked or with frozen fish for eating immediately. And what an improvement on fish fingers!

1 lb (500 g) potatoes	3 tbls (45 ml) chopped parsley
1 lb (500 g) cod or haddock fillets	salt and freshly ground pepper
1½ oz (40 g) butter, melted	dried breadcrumbs
2 small eggs	oil and butter, for frying

Boil the potatoes in salted water, drain and mash well.

Poach the fish fillets very gently in a little water or fish stock for 15 to 20 minutes until cooked. Strain, remove the skin and any bones and flake the flesh.

Mix well together the fish, potatoes, butter, eggs and parsley. Season generously. If the mixture is rather soft, chill in the refrigerator for 1 hour.

Form into fish cakes with well-floured hands, and coat with the breadcrumbs.

To serve immediately: fry gently in the oil and butter until the fish cakes are golden and hot all through – about 5 minutes on each side.

To freeze: open-freeze or wrap with clingfilm. Pack in polythene bags or plastic boxes.

To serve after freezing: fry straight from frozen, for about 10 minutes on each side.

Smoked Haddock and Sweetcorn

Serves 4

1¼ lb (600 g) frozen smoked
  haddock fillets
2½ oz (65 g) butter
2½ oz (65 g) plain flour
1½ pts (900 ml) milk, or half milk
  and half chicken, fish or veal
  stock

1 × 11-oz (325-g) can sweetcorn
  kernels
freshly grated nutmeg
salt and freshly ground pepper
2 tbls (30 ml) grated cheese
cayenne pepper or paprika
3 hard-boiled eggs

Heat the oven to 180°C, 350°F, gas 4.

Place the frozen haddock fillets in a pan, nearly cover with water and simmer gently, covered, for 15 to 20 minutes, until cooked. Alternatively, bring slowly to the boil, uncovered, remove from the heat, cover and leave. The fish will be cooked by the time it is cool enough to handle. Drain, skin and flake the flesh.

Melt the butter in a saucepan, add the flour and stir over a gentle heat for 2 to 3 minutes. Gradually stir in the milk or milk and stock until the sauce is smooth and thick. Away from the heat, stir in the fish and the strained sweetcorn. Add a little nutmeg and season. Pour into a well-buttered, shallow flameproof dish. Sprinkle with the grated cheese and a very little cayenne pepper or paprika.

Heat through in the oven for about 30 minutes, or until hot and bubbling. Just before serving sprinkle the finely chopped eggs over the top.

Note: you can also use frozen sweetcorn. Boil until just tender before adding to the sauce.

Kedgeree

Serves 4

An excellent supper dish which is not often met with nowadays. The spices make this recipe particularly tasty. The brown rice also adds flavour, as well as being healthier than polished rice; it takes longer to cook, but is worth it.

1 lb (500 g) frozen smoked
  haddock fillets
slice of lemon
a few sprigs of parsley
1 medium onion
4 oz (100 g) butter
½ tsp (2.5 ml) ground turmeric
2 cloves
2 cardamom pods or 1 tsp (5 ml)
  ground cardamom

¼ tsp (1.2 ml) caraway seeds
1-in (2.5-cm) cinnamon stick, broken
  into small pieces
8 oz (225 g) natural brown or
  long-grain rice
4 eggs
freshly ground pepper

Put the frozen haddock fillets into a large saucepan with just enough water to cover. Add the slice of lemon and the parsley, cover, bring to the boil and simmer very gently for about 10 minutes until the fish is cooked through. Alternatively, bring slowly to the boil with the pan uncovered, remove immediately from the heat, cover and leave. The fish will be cooked by the time it is cool enough to handle. Strain, reserving the cooking liquid.

Cook the finely chopped onion in half the butter until soft. Add the turmeric, cloves, cardamom, caraway seeds and cinnamon stick and stir for a minute or two. Wash the rice well and add gradually, stirring, until golden (not more than 3 minutes). Pour in 1 pt (600 ml) of the reserved cooking liquid, cover and simmer until the rice is soft but not mushy (30 to 40 minutes for brown rice, about 10 to 15 minutes for polished rice). Add a little more liquid if necessary.

Meanwhile remove the skin and any bones from the fish and flake the flesh. Hard-boil the eggs. When the rice is cooked stir in the remaining butter and fluff up with a fork. Add the fish and season to taste with freshly ground pepper. You are not likely to need salt, as the haddock will probably be salty enough. Stir over a gentle heat until the fish has warmed through.

Transfer to a heated serving dish and arrange the quartered hard-boiled eggs on top. Serve very hot.

Smoked Haddock Ramekins

Serves 4–6

These make a good starter, or a light supper dish. Arbroath smokies, the hot-smoked young haddocks from Scotland, are particularly good, but ordinary fillets of smoked haddock, or even smoked cod, may also be used.

1 lb (500 g) frozen smoked haddock or cod fillets	freshly ground pepper
½ pt (300 ml) milk	*4 tbls (60 ml) double cream*
1 onion	*finely chopped parsley*
2 egg yolks	
3 whole eggs (medium), or 2 large eggs	

Put the frozen fish fillets into a saucepan, add the milk and the quartered onion, and bring very slowly to the boil. As soon as it is boiling, remove from the heat, cover and leave until quite cool. Lift the fish out of the pan, remove the skin and any bones and flake the flesh.

Heat the oven to 180°C, 350°F, gas 4.

Mix the egg yolks with the whole eggs in a bowl and beat briefly with a fork to amalgamate.

Strain in the fish cooking liquid and stir well. Add the flaked fish and season with a little pepper (you are very unlikely to need any salt).

Pour into individual ramekin dishes. Place in a roasting tin, add cold water to come halfway up the sides of the dishes and cook in the oven for about 20 minutes, until set. Test with a knife, which should come out clean.

Top each ramekin with a little cream and a sprinkling of parsley and serve hot.

Kipper Salad

Serves 4–6

An excellent and unusual starter. Be sure to use filleted kipper, so that no one will have to struggle with bones.

2 × 8-oz (225-g) packets frozen kipper fillets	pinch of sugar
1 large Spanish onion	6–8 black peppercorns
juice of 1 large lemon	2–3 bay leaves
¼ pt (150 ml) olive or other good quality oil	*1 tbls (15 ml) roughly chopped parsley*

Allow the fillets to thaw sufficiently for you to separate and if necessary skin them, then cut diagonally into strips about 2 in (5 cm) wide. Place them in a shallow dish.

Slice the onion into fine rings. Spread over the kipper fillets.

Mix the lemon juice with the oil and sugar and pour over the kippers. Sprinkle on the roughly crushed peppercorns, top with the bay leaves and leave to marinate in the refrigerator for at least 5 to 6 hours, or overnight.

Before serving, remove the bay leaves and sprinkle with the parsley. Serve with lemon wedges and slices of brown bread and butter.

Crab au Gratin

Serves 4–6

A delicious and quite rich starter or supper dish, which can be made in about 30 minutes from start to finish.

4 oz (100 g) butter
2 oz (50 g) plain flour
¼ pt (150 ml) white wine
1 pt (600 ml) milk
1 × 1-lb (500-g) packet frozen
  crabmeat

2 tbls (30 ml) double cream
1 egg yolk
salt and freshly ground pepper
1 tbls (15 ml) fine breadcrumbs

Melt 3 oz (75 g) of the butter in a large saucepan, add the flour and cook, stirring, without allowing to brown, for 3 minutes. Add the wine and stir to a smooth paste. Slowly add the milk and stir until thick and smooth.

Add the frozen crabmeat and cook over a low heat until the crabmeat is completely thawed and blended into the sauce.

Whisk the cream into the egg yolk with a fork and stir into the sauce off the heat. Season to taste.

Spoon the mixture into six ramekin dishes or one shallow gratin dish. Sprinkle with the breadcrumbs, dot with the remaining butter and brown quickly under a hot grill.

Crab Mousse

Serves 6–8

Serve as a starter, or as part of a cold buffet.

½ pt (300 ml) single cream
1 × 1-lb (500-g) packet frozen
  crabmeat
4 eggs
2 lemons
2 tbls (30 ml) water

½ oz (15 g) powdered gelatine
6 tbls (90 ml) dry sherry
1 tsp (5 ml) cayenne
salt and freshly ground pepper

sprigs of parsley

Gently heat the cream with the frozen crabmeat until the crab has completely thawed. Bring quickly to the boil for 1 minute, then remove from the heat.

Separate the eggs and stir the yolks one by one into the crab mixture. Leave to cool. Put the juice of one lemon and the water into a small saucepan, sprinkle on the gelatine and leave for a few minutes until spongy. Heat very gently and stir until the gelatine has completely dissolved and the liquid is clear.

Stir three-quarters of the liquid gelatine into the crab mixture and add 4 tbls (60 ml) of the sherry and the cayenne. Season to taste.

Whisk the egg whites until they stand in peaks but are not dry. Fold carefully into the crab mixture, and pour into a 2-pt (1.2 L) soufflé dish.

Cut the remaining lemon into thin slices and arrange these, together with the sprigs of parsley, decoratively over the top. Add the remaining sherry to the remaining gelatine mixture, reheat gently if necessary and pour carefully over the top of the mousse.

Chill in the refrigerator for a few hours, or overnight, until set, before serving.

Crab Soufflé

Serves 4–6

An excellent starter for six, or a light but satisfying supper dish for four.

4 oz (100 g) butter
2 oz (50 g) plain flour
¾ pt (450 ml) milk
1 × 1-lb (500-g) packet frozen crabmeat
4 egg yolks
¼ pt (150 ml) double cream

2 tbls (30 ml) sherry
2 tbls (30 ml) grated parmesan or dry cheddar
salt and freshly ground pepper
5 egg whites
1 oz (25 g) flaked almonds

Heat the oven to 190°C, 375°F, gas 5.

Melt the butter in a saucepan, add the flour, stir well and cook for 2 to 3 minutes without allowing the mixture to brown. Blend the milk in slowly and stir over a gentle heat until thick and smooth.

Remove from the heat and add the frozen crabmeat. Stir until the crabmeat has thawed and is well distributed throughout the sauce, then stir in the egg yolks one by one, and add the cream, sherry and 1 tbls (15 ml) of the grated cheese. Season.

Whisk the egg whites until they stand in soft peaks and fold in gently.

Pour the mixture into a well-buttered 2-pt (1.2 L) soufflé dish, or into individual ramekin dishes, and sprinkle with the flaked almonds and the remaining cheese. Bake the large soufflé for 35 to 40 minutes, the ramekins for 15 to 20 minutes. Serve at once.

Mussels au Gratin

Serves 4–6

An excellent starter, very quickly made with frozen mussels, especially if you also have a supply of breadcrumbs in the freezer.

1 lb (500 g) frozen mussels
1 large Spanish onion
1 bunch of parsley
¼ pt (150 ml) dry white wine

½ lemon
salt and freshly ground pepper
2 oz (50 g) fine dried breadcrumbs
1 oz (25 g) butter

Allow the mussels to thaw for 30 minutes to 1 hour. They do not need to be completely thawed, but should not be cooked straight from frozen, which would make them tough.

Put the mussels, together with half the finely chopped onion, some parsley stalks and the wine, into a wide saucepan, and bring just to the boil. Remove from the heat. Remove the parsley stalks and transfer the mussels, with their juice, to a shallow ovenproof dish. Sprinkle on the lemon juice and season to taste.

Chop the parsley finely, together with a little finely pared lemon rind. Mix the breadcrumbs with the remaining chopped onion and the parsley and lemon rind and spread evenly over the mussels.

Heat the oven to 190°C, 375°F, gas 5. Dot with the butter and bake near the top of the oven for 20 minutes, until the top is just nicely browned. Serve very hot.

Seafood Pasta

Serves 4–6

Serve as a starter, or a family supper dish.

1 lb (500 g) frozen mussels	1 lb (500 g) tagliatelle or
1 small onion	fettucine noodles
1 clove garlic (optional)	1 tsp (5 ml) olive oil
small bunch of parsley	¼ pt (150 ml) double or sour cream
¼ pt (150 ml) dry white wine	salt and freshly ground pepper

Allow the mussels to thaw for 30 minutes to 1 hour. Put the mussels, together with the roughly chopped onion and garlic, some parsley stalks and the wine, into a wide saucepan and bring just to the boil. Immediately remove from the heat, cover and leave.

Boil the noodles in plenty of salted water until just cooked (*al dente*). Drain, rinse the pan, add the oil and return the pasta. Keep hot over a very low heat.

Strain the mussels and bring the juice to the boil, add the cream, stir well and boil for just one minute. Add the finely chopped parsley and the mussels and season to taste.

Pour over the noodles and serve very hot.

Prawns in Cream and Brandy Sauce

Serves 4–6

These may be served on a bed of plain boiled rice, or used to fill vol-au-vent cases or savoury pancakes (see p. 139).

1 lb (500 g) frozen peeled prawns
2 oz (50 g) butter
1 medium onion
1 tsp (5 ml) concentrated tomato
    purée
pinch of curry powder

2 tbls (30 ml) brandy
¼ pt (150 ml) dry white wine
¼ pt (150 ml) double cream
salt and freshly ground pepper

Allow the prawns to thaw for at least an hour at room temperature.

Melt the butter in a heavy frying pan or wide saucepan and soften the finely chopped onion. Add the prawns and turn over a gentle heat until they are completely thawed. Add the tomato purée and curry powder and leave to simmer very gently.

Meanwhile, heat the brandy in a small saucepan and set it alight. Douse the flames with the wine and add the cooking liquid from the prawns. Bring to the boil and boil for a few minutes until reduced by half. Add the cream and simmer until reduced to a thick sauce.

Add the prawns and onions to the sauce, season to taste and heat through well. Serve very hot.

Spanish Fried Prawns

Serves 4

This is best made with prawns frozen in their shells: they are messy but fun to eat, and have a really good flavour. A lovely way to start a meal.

1 lb (500 g) prawns frozen in
    their shells
¼ pt (150 ml) olive oil
4 cloves garlic

good pinch of cayenne pepper
freshly ground pepper

crusty French bread

Thaw the prawns thoroughly for this dish, as they must be fried quite fast.

Heat the olive oil in a deep frying pan and quickly fry the peeled and sliced garlic. As soon as it begins to colour, add the prawns and fry them very quickly in the sizzling hot oil, turning them over once. The shells should be fried crispy brown but the prawns should not be in the pan for more than 2 to 3 minutes.

Sprinkle with cayenne and black pepper. Serve immediately in small heated bowls, spooning a little of the hot garlicky oil over each serving. Accompany with good chunks of bread to mop it up.

Mediterranean Prawn Ragoût

Serves 6

A rich and substantial dish from the warm south, to be served with rice.

1 lb (500 g) peeled prawns, or 2 lb (1 kg) prawns frozen in their shells	1 red pimiento
2 large onions	2 tbls (30 ml) olive oil
2 cloves garlic	pinch of sugar
1 lb (500 g) tomatoes	salt and freshly ground pepper
1 green pimiento	½ tsp (2.5 ml) paprika

There is no need to thaw peeled prawns, but prawns frozen in their shells must be thawed sufficiently to allow them to be peeled.

Peel the onions and slice them finely into rings. Peel and chop the garlic. Skin the tomatoes and chop them roughly. Wash and seed the pimientos and cut them into strips.

Heat the oil in a flameproof casserole and soften the onion and garlic. As soon as they turn soft and yellow, add the tomatoes and increase the heat. Stir well and cook until some of the liquid from the tomatoes has evaporated. Add the pimientos, sugar, salt and pepper and paprika. Cover and simmer until the pimientos are soft.

Add the prawns and leave to thaw and heat through thoroughly over a low to moderate heat.

Check the seasoning and serve very hot on a bed of rice.

Note: if you have used prawns frozen in their shells, make a stock from the shells for prawn bisque (see p. 54).

Paprika Prawns with Ginger

Serves 4–6

A lovely spicy dish, excellent for a party, especially bearing in mind the growing number of fish-eating 'vegetarians'. It can be quickly prepared at the last minute: the only time-consuming part is peeling the prawns, but this is well worth doing as prawns frozen in their shells have much more flavour. Serve with boiled rice.

2 lb (1 kg) prawns frozen in their shells	2 tsp (10 ml) paprika
1 tbls (15 ml) oil	salt and freshly ground pepper
1 large onion	good squeeze of lemon juice
1 large clove garlic	1 tbls (15 ml) brandy or pernod
½ oz (15 g) piece of fresh root ginger	¼ pt (150 ml) double or sour cream
1 chilli pepper (optional)	1 tbls (15 ml) chopped parsley

Peel the prawns as soon as they have thawed sufficiently (keep the shells to make Prawn Bisque, see p. 54). Heat the oil in a large frying pan or wide shallow braising pan, or use a wok. Sweat the finely chopped onion and garlic until soft and add the finely chopped ginger, chilli, if you are using it, and paprika. Stir over a moderate heat until well-softened and amalgamated. Season to taste and add the lemon juice. You can prepare the dish ahead up to this point.

Shortly before serving add the brandy or pernod and bring quickly to the boil. Lower the heat, add the prawns and heat through well. Stir in the cream, add the parsley and serve very hot.

Prawn and Mushroom Vol-au-Vents

Serves 4–6

Serve these for an elegant hot hors d'oeuvre, or as a light main course. The dish can be prepared ahead without the prawns, which can be heated through just before serving.

8 oz (225 g) frozen peeled prawns	¼ pt (150 ml) double cream
1 oz (25 g) butter	½ tsp (2.5 ml) curry powder
1 large onion	salt and freshly ground pepper
8 oz (225 g) mushrooms	vol-au-vent cases (2 small or
1 tbls (15 ml) brandy	1 large per person)
1 tbls (15 ml) white wine	

There is no need to thaw the frozen prawns.

Melt the butter in a small heavy saucepan and soften the finely chopped onion. Add the sliced mushrooms and raise the heat, stirring well, so that the mushrooms cook quickly. When they are soft add the brandy and wine and stir over a high heat to evaporate most of the liquid.

Add the cream, bring to the boil and cook for a few minutes to thicken. Add the curry powder and season to taste. Shortly before serving stir in the prawns and heat through.

Heat the vol-au-vent cases for 10 minutes in a hot oven (200°C, 400°F, gas 6), then fill with the hot prawn and mushroom sauce just before serving.

Prawn and Potato au Gratin

Serves 6

Prawns add an unexpected and subtle flavour, as well as a touch of luxury, to this substantial family dish, which makes an excellent supper served with a green salad.

1 lb (450 g) frozen peeled prawns	½ pt (300 ml) chicken stock
2 lb (1 kg) potatoes	2 oz (50 g) grated cheddar or
salt and freshly ground pepper	gruyère cheese (optional)
2 oz (50 g) butter	

There is no need to thaw the frozen prawns.

Heat the oven to 190°C, 375°F, gas 5.

Peel the potatoes and slice them very thinly, using a mandolin or the slicing blade of a grater of food processor. Leave the slices in cold water for at least 10 minutes to extract the starch.

Drain and dry the potato slices, and make a thin layer on the bottom of a wide, shallow gratin dish. Sprinkle with salt and pepper, dot with a little butter and scatter on some of the prawns. Repeat until all has been used up, finishing with a layer of potatoes dotted with butter.

Bring the chicken stock to the boil and pour gently into the dish. Bake for 50 minutes to 1 hour, or until the potatoes are tender and the top layer is lightly browned, and all the liquid has been absorbed. Sprinkle with the grated cheese, if you are using it, 20 minutes before the end of the cooking time, so that it is just melted.

Prawn Bisque

It would be a pity to throw away the shells of prawns when you have made any of the previous dishes. They make stock of intense flavour, which can be stored in the freezer, ready for making this luxurious soup.

Stock

shells from at least 2 lb (1 kg) prawns   1 carrot
1 tbls (15 ml) gin or pernod   1 celery stalk
1 onion

Rinse the prawn shells and put them into a large, heavy saucepan. Add the gin or pernod and allow to evaporate for a minute over high heat. Add the roughly chopped onion, carrot and celery and cover with plenty of cold water. Bring to the boil, then leave to simmer, uncovered, for at least one hour. Bring quickly to the boil again for 1 to 2 minutes, then remove from the heat and strain. The stock should have a rich deep colour and be much evaporated, but if you wish to freeze it and there is still too much liquid leave at a rolling boil for a few minutes more before straining. Cool and freeze the stock in a plastic container if you do not wish to make the soup immediately.

Soup

1½ pts (900 ml) prawn stock   dash of gin, pernod or dry sherry
1 tbls (15 ml) plain flour   salt and freshly ground pepper
1 oz (25 g) butter
¼ pt (150 ml) double or sour cream   *a little finely chopped parsley*
1 egg yolk

Heat the stock (this may be done straight from frozen) and bring to simmering point.

Blend the flour into about half the softened butter to make a *beurre manié* and slowly whisk small pieces of this into the stock until completely amalgamated. Allow to simmer for a few minutes. Stir the cream into the egg yolk and add to the soup off the heat. Add the gin, pernod or sherry, stir in the remaining butter cut into small pieces, season to taste and serve the soup, very hot, sprinkled with the parsley.

Scallops au Gratin

Serves 2–4

This makes an excellent, delicate starter for four, or a delicious supper dish for two.

1 lb (450 g) frozen scallops	2 tbls (30 ml) finely grated gruyère,
1 onion	parmesan or cheddar cheese
¼ pt (150 ml) dry white wine or cider	4 tbls (60 ml) double cream
½ oz (15 g) plain flour	salt and freshly ground pepper
2 oz (50 g) butter	2 tbls (30 ml) fine dry breadcrumbs

Allow at least 30 minutes for the scallops to thaw, or at any rate part-thaw, before cooking.

Put the scallops into a saucepan with the finely chopped onion, the wine or cider and enough water to just cover. Heat slowly, then poach for 6 to 8 minutes or until the scallops are opaque and tender. Remove from the pan with a slotted spoon, slice thickly and set aside.

Boil the cooking liquid briskly for 2 minutes to reduce, then strain and return to the pan. Make a *beurre manié* by working the flour into 1½ oz (40 g) of the butter. Slowly whisk small pieces into the liquid and simmer for 2 minutes. Add half the cheese and the cream and season to taste.

Return the scallops to the sauce and heat through, then pour into heated ramekin dishes or into a 1-pt (600-ml) shallow ovenproof dish. Sprinkle with the breadcrumbs mixed with the remaining cheese, dot with the remaining butter and brown quickly under a hot grill.

Shrimp Hors d'Oeuvre

Serves 4

A light and unusual starter.

6 oz (175 g) frozen peeled shrimps or prawns	1 boursin cheese
1 oz (25 g) flaked almonds	1 tbls (15 ml) single cream

Heat the oven to 150°C, 300°F, gas 2.

Allow the shrimps or prawns to thaw and pour off any liquid. Toast the almonds to a light brown.

Gently warm the cheese in a small saucepan, gradually stirring in the cream until the mixture is smooth. Add the shrimps or prawns. Divide between four ramekins and top with the flaked almonds.

Cover loosely with foil and heat through in the oven for 30 to 40 minutes.

Shrimp and Walnut Hors d'Oeuvre

Serves 4

A very refreshing appetizer which can be made at short notice, as peeled shrimps or prawns do not take long to thaw.

4 oz (100 g) frozen peeled shrimps or prawns	good squeeze of lemon juice
	salt and freshly ground pepper
1 small lettuce	½ tsp (2.5 ml) French mustard
2 eating apples	2 tbls (30 ml) double or sour cream
4 oz (100 g) walnuts	

Allow the shrimps or prawns to thaw. Chop prawns roughly if you wish, keeping a few whole for decoration.

Wash the lettuce and line a salad bowl with the leaves. Peel, core and roughly chop the apples. Chop the walnuts. Mix shrimps, apples and walnuts in a bowl, and sprinkle with the lemon juice. Season to taste.

Stir the mustard into the cream, add a little more seasoning and fold into the apple, shrimp and walnut mixture.

Spoon into the centre of the salad bowl and serve.

Pasta

∗∗∗∗∗∗∗∗∗∗∗∗∗∗∗∗∗∗

Making pasta at home is fun if you have plenty of time, lots of patience and a pasta-making machine. Otherwise it is better to rely on the increasing number of shops which nowadays sell fresh pasta in a variety of shapes, tagliatelle, taglierini, ravioli, lasagne and tortellini being the most common. Fresh pasta is so infinitely superior to the dried variety that it is worth making an effort to find a supply and freezing it, as it keeps well in the freezer for a month or two.

Frozen pasta should be cooked straight from the freezer, and all pasta should be cooked in the largest possible quantity of salted water. Bring the water to the boil and put in the pasta, stir with a fork so that it doesn't stick together, and cover the pan until the water comes to the boil again. Take off the lid and boil until the pasta is still slightly resistant to the teeth (*al dente*). With fresh pasta, even if straight from the freezer, this will take only a few minutes, so be careful not to overcook it, as soft pasta is horrible. Taglierini, in fact, will probably be ready as soon as the water has come to the boil again.

The moment the pasta is cooked, strain into a colander and shake a few times to get rid of excess water. Return to the pan and either mix with the sauce or stir in a little butter or oil to prevent the pasta from sticking.

Some of the most popular sauces can also be frozen, so with both pasta and sauce in the freezer an appetizing meal can be prepared very quickly. Freeze the sauce in separate containers in the quantities you are likely to need.

Parmesan is always the best cheese to have with pasta dishes, but it is very expensive and a mature cheddar will often do nearly as well. If you do use parmesan, make sure to buy it in a piece for grating yourself, since the ready-grated parmesan, bought in small containers, looks and tastes rather like sawdust.

SAUCES

The three sauces given below all freeze well.

Tomato Sauce

This sauce is quick to prepare, and has a very fresh taste. It will keep indefinitely in the freezer, and for at least a month in the refrigerator.

1 tbls (15 ml) olive oil	2 or 3 sprigs of parsley
1 medium onion	2 lb (1 kg) ripe tomatoes
1 celery stalk	salt and freshly ground pepper
1 clove garlic	sugar
1 bay leaf	a few basil leaves (optional)
sprig of thyme	

Heat the olive oil in a saucepan and cook the chopped onion and celery, the crushed garlic, the bay leaf, thyme and parsley over a fairly brisk heat, stirring from time to time. When the onion is golden add the roughly chopped tomatoes. Season with salt, pepper and a little sugar, cover and simmer for 20 minutes. Add the basil, if you are using it, towards the end of the cooking time. Remove the bay leaf and thyme, pour into a blender or food processor and blend, then sieve. Check the seasoning.

To serve immediately: return to the pan and reheat.

To freeze: allow to cool, then freeze in well-sealed disposable cream or yoghurt pots, as the flavour tends to linger.

To serve after freezing: tip the frozen sauce into a saucepan and reheat very gently.

Note: a 1¾-lb (795-g) can of tomatoes can be used for this sauce instead of fresh ones. You will need only about half the juice, and do not cover the pan.

Meat Sauce

1 onion
1 celery stalk
3–4 basil leaves (optional)
2 tbls (30 ml) olive oil
1 lb (500 g) minced beef
1 lb (500 g) fresh tomatoes or
  1 × 14-oz (400-g) can tomatoes

1 tbls (15 ml) concentrated tomato
  purée
1 tsp (5 ml) dried mixed herbs
salt and freshly ground pepper

Sauté the sliced onion, celery stalk and basil, if you are using it, in the olive oil in a saucepan until the onion is golden. Add the minced beef and cook for a few minutes, stirring, until it changes colour.

If you are using fresh tomatoes, skin and chop roughly. If you are using canned, sieve them together with the juice. Add to the pan, with the tomato purée and the dried herbs, and season. Cover and simmer gently for about 1 hour, stirring from time to time, until the sauce is thickened and reduced.

If necessary, leave the lid off the pan for the last few minutes of cooking, if the sauce hasn't thickened enough.

To serve immediately: return to the pan and reheat.

To freeze: allow to cool, then freeze in well-sealed disposable cream or yoghurt pots, as the flavour tends to linger.

To serve after freezing: tip the frozen sauce into a saucepan and reheat very gently.

Pesto Genovese

One of the classic and most fragrant Italian sauces, which, since basil is rarely obtainable in shops, can normally be made only if you grow it yourself. It is well worth keeping a few basil plants in pots throughout the summer and autumn, or freezing the leaves. Only the young leaves should be used.

Ideally the cheese used should be a mixture of parmesan and pecorino. If you can't find pecorino, use all parmesan.

Pesto genovese can be made quickly and easily in a blender or food processor. As well as providing a delicious sauce for pasta, a spoonful or two added to stews and soups imparts a subtle and unusual flavour.

2 cloves garlic

3 oz (75 g) young basil leaves,
weighed after stripping from
the stalks

½ pt (300 ml) best olive oil

2 oz (50 g) pine kernels

2 oz (50 g) grated cheese, either all
parmesan or half parmesan and
half pecorino

salt

Crush the garlic with the flat blade of a knife and slip off the skin. Put the garlic, basil, olive oil and pine kernels into a blender or food processor and blend at high speed until the sauce is absolutely smooth.

To serve immediately: stir in the grated cheese and season to taste. If you are going to freeze the sauce, do not add the cheese and salt until after it has thawed.

To freeze: freeze in small, well-sealed pots.

To serve after freezing: thaw overnight in the refrigerator.

Pesto will also keep well in the refrigerator for several weeks, so long as there is a film of olive oil on top.

Tagliatelle with Cream

Serves 6–8 as a starter, 4–6 as a main course

One of the most delicious pasta dishes of all, rich but irresistible.

1 lb (500 g) white or green tagliatelle

4 oz (100 g) butter

2 oz (50 g) freshly grated parmesan
cheese

freshly ground pepper

½ pt (300 ml) double cream

Frozen tagliatelle can be cooked straight from the freezer.

Cook the tagliatelle in plenty of boiling salted water (see p. 57). As soon as it is done strain through a colander, shake to remove any excess water and return to the pan.

Over a low heat, mix the butter in well. Add the cheese and lots of pepper. Finally stir in the cream and serve at once.

'Hay and Straw' (Fieno e Paglia)

Serves 6–8 as a starter, 4–5 as a main course

This dish is made with a mixture of green and white taglierini, and is best served very simply with lots of butter and grated cheese, but a light tomato sauce also goes well with it. The contrast between the colours is very attractive.

8 oz (225 g) white taglierini	freshly ground pepper
8 oz (225 g) green taglierini	freshly grated parmesan cheese
butter	

Frozen taglierini can be cooked straight from the freezer.

Cook the taglierini in a large saucepan of boiling salted water (see p. 57).

Strain through a colander, return to the pan, and mix quickly with enough butter to prevent it from sticking. Season generously with pepper. Serve very hot, handing round more butter and grated parmesan cheese.

Tagliatelle Primavera

Serves 6–8

A refreshing vegetarian supper dish to make in early summer, which not only looks pretty but is simple and quick to make in large quantities.

1 lb (500 g) tagliatelle or taglierini, preferably white	8 oz (225 g) button mushrooms
½ oz (15 g) butter	½ pt (300 ml) double cream
small bunch of spring onions	4 oz (100 g) grated cheese, preferably parmesan
1 red pepper	salt and freshly ground pepper
8 oz (225 g) mange-tout peas	

Frozen tagliatelle can be used straight from the freezer.

Melt the butter in a large, deep frying pan or a wok. Add the trimmed and chopped onions and cook gently until just soft. Add the pepper, washed, cored and seeded and cut into strips, and turn over moderate heat until it too just begins to soften.

Add the mange-tout peas, washed and stringed, and the thickly sliced mushrooms and cook for 2 to 3 minutes, until the peas have changed colour but are still quite crisp.

Cook the pasta in plenty of boiling salted water (see p. 57). As soon as it is done, drain and add to the vegetables in the pan. Heat through together gently but thoroughly.

Bring the cream to the boil in a small saucepan, pour over the dish, add half the grated cheese and plenty of salt and pepper, and toss gently.

Serve very hot, handing round the remaining cheese separately.

Variation: this is also delicious if you add one or two chicken breasts. Skin and bone them, and cut diagonally into thin slices. Add to the pan with the red pepper.

You can, if you wish, marinate the chicken slices in a mixture of soy sauce and sherry for an hour or two first, to give extra flavour.

Tagliatelle alla Carbonara

Serves 6–8 as a starter, 4–6 as a main course

A classic way of serving pasta, excellent either as a starter or as a substantial main course. Spaghetti is normally used, but tagliatelle is equally good.

1 lb (500 g) tagliatelle
8 oz (225 g) lean bacon rashers
1 tbls (15 ml) olive oil
4 eggs
4 oz (100 g) grated parmesan or
    mature cheddar cheese

3 tbls (45 ml) cream
2 oz (50 g) butter
freshly ground pepper

Frozen tagliatelle can be used straight from the freezer.

Remove the rind from the bacon. Dice the bacon and fry gently in the oil for a few minutes. Set aside.

Beat the eggs and mix in the grated parmesan and the cream.

Cook the tagliatelle in plenty of boiling salted water (see p. 57). As soon as it is done, strain and return to the pan. Over a low heat mix in the butter, and quickly stir in the bacon and the egg mixture. Season with plenty of pepper.

The moment everything is very hot, serve at once on well-warmed plates.

Tagliatelle with Red Peppers and Prosciutto Ham

Serves 6–8 as a starter, 4–6 as a main course

1 lb (500 g) tagliatelle
2 medium red peppers
3 tbls (45 ml) olive oil
6 oz (175 g) prosciutto ham, cut in
   one slice
6 oz (175 g) frozen peas

freshly ground pepper
⅓ pt (200 ml) double cream
3 oz (75 g) grated cheese, preferably
   parmesan

Frozen tagliatelle can be cooked straight from the freezer.

Core and seed the peppers and cut into very small dice. Cook gently in the oil, in a covered saucepan, until quite soft.

Add the ham, also cut into small dice, and the peas (no need to thaw them). Season with pepper and simmer for a few minutes. As the prosciutto will be salty, there is no need to add salt.

Stir in the cream and allow to bubble over a fairly high heat, stirring from time to time, until it has reduced and the sauce is thick and creamy.

Stir in the grated cheese, remove from the heat and keep warm while you cook the pasta.

Cook the tagliatelle in plenty of boiling salted water (see p. 57). As soon as it is done strain through a colander, shake to remove any excess water, and return to the pan. Stir in the sauce and serve at once.

Tagliatelle Pie with Chicken Livers and Ham

Serves 6–8 as a starter, 4–6 as a main course

10 oz (275 g) tagliatelle
1 large onion
4 oz (100 g) butter
4 oz (100 g) mushrooms
8 oz (225 g) chicken livers
8 oz (225 g) cooked ham

salt and freshly ground pepper
freshly grated nutmeg
2 oz (50 g) fresh breadcrumbs
2 eggs
¼ pt (150 ml) single cream
3 oz (75 g) grated cheese

Heat the oven to 180°C, 350°F, gas 4.

Frozen tagliatelle can be used straight from the freezer.

Cook the finely chopped onion gently in half the butter in a large frying pan until soft and golden. Add the finely chopped mushrooms and fry for a further 3 to 4 minutes. Chop the chicken livers, discarding any fat or green bits, and add to the pan, stirring over a low heat until the livers are cooked but still slightly pink inside. Add the diced ham and cook for a further 4 to 5 minutes.

Boil the tagliatelle in plenty of salted water (see p. 57). As soon as it is done strain, and off the stove mix at once with the ham and liver mixture. Season and add a little nutmeg.

Butter an ovenproof dish and sprinkle in half the breadcrumbs. Put in the tagliatelle mixture. Beat the eggs with the cream, add the cheese, season with a little pepper and pour over the tagliatelle. Sprinkle on the remaining breadcrumbs and dot with the rest of the butter.

Bake uncovered for about 30 minutes, until the pie is golden and bubbling.

Tagliatelle with Olives and Basil

Serves 6–7

A starter for the summer, when tomatoes and basil are at their best. It is best eaten without grated cheese, in order not to spoil the fresh taste of the tomatoes and basil.

1 lb (500 g) tagliatelle	2 tbls (30 ml) olive oil
1 lb (500 g) very ripe tomatoes	salt and freshly ground pepper
about 24 black olives	2 oz (50 g) butter
about 20 basil leaves	

Frozen tagliatelle can be used straight from the freezer.

Skin and halve the tomatoes, remove the seeds and leave upside down to drain. Halve the olives and take out the stones. Chop the basil (not too finely), and mix with the olives, the olive oil and the roughly chopped tomatoes. Season, but be careful with the salt, as the olives will be salty.

Cook the tagliatelle in plenty of boiling salted water (see p. 57), and when it is done, drain, return to the pan and stir in the butter over a low heat. As soon as the tagliatelle is coated with the butter remove from the heat and quickly stir in the sauce. Serve at once.

Taglierini Ring with Mushrooms and Peas

Serves 6–8 as a starter, 4–6 as a main course

1 lb (500 g) taglierini
8 oz (225 g) mushrooms
about 6 oz (175 g) butter
a few sprigs of parsley
1 clove garlic
salt and freshly ground pepper

a little sugar
8 oz (225 g) young shelled fresh or
  frozen peas

freshly grated parmesan cheese

Frozen taglierini can be used straight from the freezer.

Heat the oven to 200°C, 400°F, gas 6.

Cook the sliced mushrooms gently in about 2 oz (50 g) of the butter, with the parsley, crushed garlic and a little salt and pepper. When the mushrooms are soft, add the peas and 1 tsp (5 ml) or so of sugar and continue to cook until the peas are tender. Keep warm.

Cook the taglierini in plenty of boiling salted water (see p. 57). The moment they are done drain, return to the pan and mix with the remaining butter and plenty of pepper. Turn into a buttered savarin mould and put in the hot oven for 10 minutes.

Unmould on to a warm dish, and fill the centre with the mushrooms and peas. Serve at once, handing round the grated parmesan.

Lasagne with Four Cheeses

Serves 6

This is a pleasant variation on the more usual lasagne with meat sauce, and is useful if you want to serve a vegetarian dish. It is not difficult nowadays to find the cheeses you need, since membership of the Common Market means that there is a far wider variety of Italian cheese in the shops than there used to be.

7 oz (200 g) green lasagne sheets
1 tbls (15 ml) salad oil

Sauce
3 oz (75 g) parmesan cheese
3 oz (75 g) gruyère cheese
3 oz (75 g) mozzarella or
  bel paese cheese

3 oz (75 g) pecorino cheese
2 oz (50 g) butter
2 oz (50 g) plain flour
1¼ pts (750 ml) milk
salt and freshly grated pepper
freshly grated nutmeg

Frozen lasagne can be cooked straight from the freezer.

Fill your largest saucepan with salted water. Add the salad oil, which will help to prevent the lasagne from sticking together. When the water is boiling hard drop in the lasagne leaves about six at a time and boil over a brisk heat, uncovered, until they are just cooked. Lift out each batch carefully and plunge at once into cold water for a few minutes. When all the lasagne is cooked, drain and lay on kitchen paper for about 30 minutes to dry.

To make the sauce, grate the parmesan and cut the other cheeses into small dice. Put about one-third of the parmesan on one side and mix all the cheeses together.

Melt the butter in a saucepan. Add the flour and stir over a low heat for 2 to 3 minutes. Add the milk gradually, stirring until the sauce has come to the boil and is thick and smooth. Add the cheeses (except the reserved parmesan). Season with salt and pepper and generously with nutmeg, and leave over a low heat, stirring occasionally, until the cheeses have melted.

Heat the oven to 180°C, 350°F, gas 4.

Generously butter a big, fairly shallow ovenproof dish and put in alternate layers of lasagne and sauce. There should be at least four layers. Finish with a layer of sauce, and sprinkle the reserved parmesan on top.

Bake for 45 minutes to 1 hour, until the sauce is bubbling and the top is golden brown.

Rice

✳✳✳✳✳✳✳✳✳✳✳✳✳✳✳✳✳✳

It is often a boon to have a few packets of cooked rice in the freezer, for using in a variety of dishes and salads or for adding to soups and casseroles, stuffing vegetables or poultry, or accompanying Chinese or Indian takeaway meals. It takes no longer to cook 2 lb (1 kg) of rice than it does 1 lb (500 g), and the extra amount can quickly be frozen. Or if you have any left over from a meal, which so frequently happens, freeze what you don't eat.

Natural unpolished brown rice, which is much healthier than white rice, and has such a delicious flavour and nutty texture, is particularly worth freezing in this way; it takes longer to cook than white, so it makes a lot of sense to cook a double quantity, half for eating at once and half for freezing.

Cook the rice, white or brown, in plenty of boiling salted water, being careful not to overcook – it should still be slightly resistant to the teeth when you take it off the stove. Drain, and run immediately under plenty of cold water. Drain well again and freeze in small quantities, so that it will thaw more quickly.

Brown rice will roughly double its weight during cooking. White rice, which absorbs more water, will produce rather more than this, so that, for instance, 6 oz (175 g) uncooked white rice will weigh nearly 1 lb (450 g) when cooked. Since brown rice absorbs less water it takes much less time to thaw than white, so it can be used very soon after you take it out of the freezer.

Risotto al Pollo

Serves 8

1 lb (450 g) cooked rice
1 lb (450 g) cooked chicken or turkey
6 oz (175 g) butter
3 oz (75 g) plain flour
2 pts (1.2 L) milk, or half milk and
    half chicken stock

8 oz (225 g) well-flavoured grated
    cheese
salt and freshly ground pepper
freshly grated nutmeg

Frozen rice and chicken or turkey should be allowed to thaw.

Heat the oven to 200°C, 400°F, gas 6.

Spread the rice over the bottom of a large, fairly shallow ovenproof dish in a layer no more than ½ in (1 cm) thick.

Cut the chicken into bite-sized pieces and arrange evenly over the rice.

Melt half the butter in a saucepan, add the flour and stir for 2 to 3 minutes over a low heat. Gradually add the milk, or milk and chicken stock, stirring until the sauce has come to the boil and is thick and smooth. Stir in half the cheese and season with salt, pepper and nutmeg. The flavour of the nutmeg will become less marked during the cooking, so add it generously.

Pour the sauce evenly over the chicken and sprinkle the remaining cheese on top. Dot with the remaining butter and cook for 30 to 40 minutes, until the top is bubbling and golden brown.

Rice and Chicken Ring

Serves 8

A good dish for a party, since it keeps warm quite happily until you are ready to eat it.

1½ lb (675 g) cooked rice
1 lb (450 g) cooked chicken, chopped
    or coarsely minced
4–6 oz (100–175 g) butter
salt and freshly ground pepper
8 tomatoes
a little sugar

2 medium onions, finely chopped
2 oz (50 g) plain flour
1 pt (600 ml) milk, or half milk and
    half chicken stock
2 oz (50 g) grated cheese
8 oz (225 g) mushrooms
¼ pt (150 ml) double or
    whipping cream

Frozen rice and chicken should be allowed to thaw.

Heat the oven to 180°C, 350°F, gas 4.

Slice the mushrooms and cook them gently in 2 oz (50 g) of the butter for about 5 minutes, until the liquid has been absorbed. Mix with the rice, season, and spoon into a well-buttered ring mould. Cover with foil. Place the mould in a roasting tin, pour in hot water to come about 1 in (2.5 cm) up the sides of the mould, and heat through in the oven for about 45 minutes.

While the rice is warming, cut the tomatoes in half, arrange them in a shallow baking tin, put a dab of butter on each, and season with a little salt and pepper and a sprinkling of sugar. Cook them in the oven with the rice.

Melt the remaining butter in a saucepan, add the onions and cook gently until soft but not brown. Add the flour and stir over a low heat for 2 to 3 minutes. Stir in the milk, or milk and chicken stock, and bring to the boil, stirring until the sauce is thick and smooth. Add the cheese, chicken and cream, and season. Leave over a very low heat, stirring from time to time, until the mixture has warmed through.

Turn out the rice and mushrooms on to a big, warmed serving dish. Pile the chicken mixture in the middle, and surround with the tomatoes.

Savoury Rice

Serves 8

A pretty, tasty dish to serve with any kind of casserole.

2¼ lb (1 kg) cooked rice	about 2 oz (50 g) butter
1 × 7-oz (200-g) can red peppers	salt and freshly ground pepper
1 × 14-oz (400-g) can petits pois	juice of ½ lemon
8 oz (225 g) mushrooms	

Frozen rice should be allowed to thaw.

Heat the oven to 150°C, 300°F, gas 2.

Drain the peppers and reserve the juice from the can. Carefully remove any seeds and slice the peppers thinly. Drain the peas. Cook the sliced mushrooms gently in half the butter, in a covered pan, for 5 to 10 minutes.

Mix together the rice, peppers, peas, mushrooms and the juices from the peppers and mushrooms. Season and put into a shallow casserole.

Dot with the remaining butter and cover with a lid or foil. Warm through in the oven for 45 minutes to 1 hour, turning with a fork from time to time.

Just before serving squeeze the lemon juice over the top.

Vegetable Pie

Serves 6–8

6 oz (175 g) cooked rice
3 medium leeks
4 oz (100 g) fresh shelled or
    frozen peas
6 eggs, separated

½ oz (15 g) butter, melted
3 oz (75 g) grated cheddar cheese
¼ pt (150 ml) sour cream
salt and freshly ground pepper

Frozen rice should be allowed to thaw.

Heat the oven to 180°C, 350°F, gas 4.

Chop the leeks into ½-in (1-cm) lengths and cook lightly in a little water. Cook the peas separately until just tender. Drain both well.

Beat the egg yolks with the butter for 5 minutes. Add the cheese and the sour cream. Whisk the egg whites until stiff but not dry and fold in. Fold in the peas, leeks and rice. Season.

Turn into a large buttered soufflé dish. Bake for 30 to 40 minutes, until the pie is golden-brown on top and just firm to the touch. Serve at once.

Rice with Four Cheeses and Mushroom Sauce

Serves 4

about 12 oz (350 g) cooked rice
1 small onion
2 oz (50 g) butter
8 oz (225 g) mushrooms
salt and freshly ground pepper
2 oz (50 g) gruyère cheese

2 oz (50 g) fontina or
    Dutch edam cheese
2 oz (50 g) bel paese cheese
1 oz (25 g) grated parmesan cheese
1 tbls (15 ml) double cream

Frozen rice should be allowed to thaw.

Heat the oven to 180°C, 350°F, gas 4.

Cook the finely chopped onion gently in half the butter in a frying pan until soft and golden. Add the sliced mushrooms, season, cover and cook over a low heat for a further 5 minutes or so.

Grate the gruyère and the fontina or Dutch edam, and cut the bel paese into small pieces. Oil or butter a ring mould, and put in a layer of rice. Dot it with a little of the remaining butter and sprinkle with the grated parmesan. Next add all the bel paese, then another layer of rice, a little more butter and the gruyère. Repeat with the fontina or Dutch edam, and finish with a layer of rice.

Place the ring mould in a roasting tin, pour in hot water to come halfway up the sides of the mould, cover loosely with foil and heat through in the oven for about 45 minutes. Stir the cream into the mushroom mixture and check the seasoning. Turn the rice mould out on to a heated serving dish and pour the mushroom mixture into the centre and all round it.

Variation: young, tender peas make a good alternative to the mushrooms. Cook them in a very little water, with salt, butter, mint and a little sugar. Or a mixture of peas and mushrooms is good, adding colour as well as taste.

Offal

✳✳✳✳✳✳✳✳✳✳✳✳✳✳✳✳✳✳

Despite its rather depressing name, offal is a delicious, relatively inexpensive form of meat, packed with nutritional value, and full of highly desirable vitamins and minerals.

Frozen packs are available in most supermarkets, and since offal comes in small pieces and therefore thaws and cooks quite quickly, a pack or two of each type kept in the freezer ensures that a good meal for family or unexpected guests can always be made at quite short notice.

LAMB'S, CALF'S AND PIG'S LIVERS

Lamb's and calf's liver has a more delicate taste than pig's, but pig's liver is excellent used in pâté.

All these livers can be used as soon as they have thawed sufficiently to be easily sliced – in fact, they are best sliced while still slightly frozen – and can be made into a great variety of tasty, economical dishes.

On the whole, it is best not to freeze liver once it has been cooked (with the exception of pâtés), and not to keep the frozen uncooked liver longer than 3 months in the freezer.

Liver with Onions

Serves 4

1 lb (500 g) frozen liver
1 tbls (15 ml) plain flour
1 lb (500 g) onions

4 oz (100 g) butter
salt and freshly ground pepper
1 tbls (15 ml) wine vinegar

Cut the liver into thick slices as soon as it has thawed sufficiently, and dust each slice lightly with the flour. Slice the onions very finely.

Melt half the butter in a frying pan, and when it is foaming and just beginning to brown add the liver and brown quickly on both sides. Remove to a heated serving dish and keep hot.

Heat the remaining butter in the pan and cook the onions over a brisk heat until browned, stirring from time to time so that they brown evenly. Season to taste, then add the vinegar and bring quickly to the boil.

Pile the onions over the liver and serve very hot.

Liver Stroganoff

Serves 4

This delicate dish is best made with calf's or lamb's liver.

1 lb (500 g) frozen liver	1 tbls (15 ml) oil
8 oz (225 g) onions	dash of wine vinegar
rind of ½ lemon	¼ pt (150 ml) sour cream
small bunch of parsley	salt and freshly ground pepper
2 oz (50 g) butter	

Cut the liver while it is still slightly frozen into 3–4 × 1-in (8–10 × 2.5-cm) strips, about ¼ in (5 mm) thick, rather as you would cut beef for Boeuf Stroganoff.

Chop the onions very finely, and then chop the lemon rind and parsley together finely.

Melt the butter and oil in a deep frying pan or flameproof casserole and cook the liver strips over a moderate heat, turning them over as soon as beads of blood begin to appear on the top side. Remove the liver as soon as it is cooked and keep warm.

Sweat the onions in the pan until soft and golden. Add to the pan the juice from the liver, together with the vinegar. Stir in the sour cream, lemon rind and parsley. Season to taste, return the liver to the pan, allow to heat through and serve very hot with creamy mashed potatoes or rice.

Liver with Sour Cream

Serves 4

1 lb (500 g) frozen liver
a little plain flour
2 tbls (30 ml) oil
1 oz (25 g) butter

¼ pt (150 ml) dry white wine
¼ pt (150 ml) sour cream
salt and freshly ground pepper
3–4 tbls (45–60 ml) chopped parsley

Cut the liver into slices while it is still slightly frozen. Coat lightly in flour. Heat the oil and butter in a large, heavy frying pan and fry the liver quickly for about 1 minute on each side: it should still be pink inside. Remove to a heated serving dish and keep warm.

Add the wine to the pan and cook for a few minutes to reduce. Add the sour cream and continue to cook for about 2 minutes more, stirring well to scrape up any brown bits in the pan to enrich the sauce. Season, and stir in the parsley. Pour the sauce over the liver and serve at once.

Liver and Mushroom Casserole

Serves 4

This is equally good for family supper or an informal dinner party. If you want to make it beforehand, it heats up very satisfactorily. It is an exception to the general rule for cooked offal, since it freezes very well; but only freeze it if you have made it with fresh, not frozen, liver.

1 lb (500 g) liver
1 lb (500 g) onions
3 oz (75 g) dripping or 2 oz (50 g)
  butter and 2 tbls (30 ml) oil

salt and freshly ground pepper
1 oz (25 g) plain flour
12 oz (350 g) mushrooms
½ pt (300 ml) beef stock

Heat the oven to 150°C, 300°F, gas 2.

Slice the liver about ⅓ in (6 mm) thick.

Slice the onions fairly finely and sweat gently in half the dripping or butter and half the oil in a large, heavy covered frying pan, until soft and golden.

Coat the liver in the well-seasoned flour. Remove the onions from the pan, turn up the heat, add the remaining dripping or butter and oil, and fry the liver quickly on both sides to seal. Remove from the pan, reduce the heat and add the sliced mushrooms. Toss for a few minutes, if necessary adding a little more fat.

Return the onions and liver to the pan, add the stock and bring to a simmer, scraping up any brown bits in the pan to add to the flavour of the gravy. Transfer to a casserole, cover and cook in the oven for 1¼ hours.

To serve immediately: serve hot with mashed potatoes.

To freeze: allow to cool, then freeze.

To serve after freezing: thaw overnight in the refrigerator or for 5 to 6 hours at room temperature. Reheat gently either in a cool oven or on top of the stove.

Braised Liver with Vegetables

Serves 4

1 lb (500 g) frozen liver	1–1½ oz (25–40 g) butter or chicken
1 red pimiento	fat
1 carrot	salt and freshly ground pepper
2 medium onions	1–2 tbls (15–30 ml) seasoned flour
2 celery stalks	6 tbls (90 ml) dry white wine
2–3 tbls (30–45 ml) oil	6 tbls (90 ml) stock

Slice the liver finely while it is still slightly frozen.

Seed the pimiento and chop very finely together with the carrot, onions and celery.

Melt the oil and the butter or chicken fat in a large frying pan. Add the vegetables, season lightly and sweat, covered, for 10 to 15 minutes until soft, stirring from time to time.

Arrange the vegetables round the edge of a heated serving dish and keep warm.

Coat the liver lightly in the seasoned flour.

Raise the heat under the frying pan and quickly cook the liver for 30 seconds on each side, adding more fat if necessary. Add the wine and stock, cover and simmer for about 15 minutes until the liver is cooked.

Transfer the liver to the serving dish, in the centre of the vegetables. Serve at once.

Liver with Orange

Serves 4

1 lb (500 g) frozen liver
1 large onion
2 oz (50 g) butter
3–4 tbls (45–60 ml) oil
1–2 tbls (15–30 ml) plain flour

2 tbls (30 ml) white wine
juice of 2 oranges
salt and freshly ground pepper

chopped parsley

Cut the liver into very thin slices while it is still slightly frozen.

Cook the finely chopped onion gently in the butter and oil in a frying pan until soft. Remove to a heated serving dish and keep warm.

Dust the liver lightly with the flour and fry very quickly, for no more than 1 minute on each side – it should still be pink in the middle. Arrange the liver on the serving dish and keep warm.

Deglaze the pan with the wine, add the orange juice and boil for a couple of minutes to reduce a little. Season lightly, pour over the liver and garnish with the parsley. Serve at once.

Liver with Sage

Serves 4

The sage for this simple but excellent dish must be young and tender, so pick the leaves from the tips of the plant.

1 lb (500 g) frozen liver
2–3 tbls (30–45 ml) oil
1 oz (25 g) butter
about 30 young, small sage leaves
2–3 tbls (30–45 ml) dry white wine

squeeze of lemon juice
salt and freshly ground pepper

slices of lemon
sprigs of parsley

Cut the liver into very thin slices while it is still slightly frozen.

Heat the oil and butter in a large frying pan and very gently cook the sage leaves; do not allow them to become dry or brown.

When you are ready to cook the liver push the sage to the edge of the pan and turn up the heat. Cook the liver quickly – about 1 minute on each side, as it should still be pink in the middle.

Transfer the liver and sage to a heated serving dish and keep warm.

Add the wine and lemon juice to the pan and season lightly. Cook briskly until reduced by about half. Pour over the liver, garnish with the lemon slices and parsley and serve at once.

Liver with Tomato Sauce

Serves 4

1 lb (500 g) frozen liver	sprig of sage
1 oz (25 g) butter	¾ pt (450 ml) tomato sauce
2 tbls (30 ml) oil	(see p. 58)
2 cloves garlic	

Slice the liver thinly while it is still slightly frozen.

Heat the butter and oil in a large, heavy frying pan and gently fry the crushed garlic and the sage until brown. Discard. Add the liver to the pan and fry briskly for no more than 1 minute on each side as it should still be pink in the middle.

Arrange the liver on a heated serving dish and pour the hot tomato sauce over the top. Serve at once.

Liver Pilaff

Serves 4

A savoury and economical dish which heats up excellently. It can also be made with chicken livers, but they must be allowed to thaw first if frozen.

12 oz–1 lb (350–500 g) frozen liver	8 oz (225 g) white long-grain or
1 medium onion	Italian risotto rice
2 medium tomatoes	about 1¾ pts (1 L) beef stock
1 small pimiento, preferably red	salt and freshly ground pepper
2 oz (50 g) butter	
2 tbls (30 ml) sultanas	*chopped parsley*

Cut the liver into bite-sized squares. This is best done while it is still slightly frozen.

Chop the onion finely. Skin the tomatoes, remove the seeds and chop finely. Seed the pimiento and slice finely.

Melt half the butter in a large, deep frying pan or flameproof casserole and cook the liver quickly for 2 or 3 minutes until it is no longer pink. Remove from the pan.

Melt the remaining butter and cook the onions and pimiento until soft but not brown. Add the tomatoes and sultanas and cook for a few more minutes. Add the rice and stir over a low heat for not more than 3 minutes. Pour in the hot stock, season and cook over a brisk heat, uncovered, for about 5 minutes, until the rice begins to absorb the stock. Place the liver on top, turn down the heat, cover and cook very gently until the rice is just tender but not mushy.

Turn on to a heated serving dish and serve at once, sprinkled with parsley.

CHICKEN LIVERS

No freezer should be without one or two tubs of frozen chicken livers, which are invaluable not only for making pâtés and adding to stuffings, but for all kinds of quick-to-make family dishes and special starters.

Chicken livers should always be allowed to thaw completely before cooking, as otherwise they will not cook evenly. An 8-oz (225-g) tub takes approximately 3 hours to thaw at room temperature or overnight in the refrigerator. Leave in the closed container to thaw, and, once thawed, use as quickly as possible.

Pick the livers over carefully before cooking – just occasionally you may find one with a greenish tinge which should be removed, as it will make the rest bitter. It is best also to trim off any bits of membrane before cooking.

Chicken Livers with Grapes

Serves 6 as a starter, 3–4 as a main course

An unusual starter, or a luxurious but light supper dish, this can be made richer still by substituting one or two duck or turkey livers. It is also an ideal quick supper dish to make a day or two before Christmas if you are having a goose, as goose liver is the richest of them all, and should not be wasted by adding to a stuffing.

1 lb (500 g) frozen poultry livers
8 oz (225 g) white grapes
6 oz (175 g) butter
1 tbls (15 ml) oil

4–6 slices of white bread
salt and freshly ground pepper
¼ pt (150 ml) dry white wine or
    vermouth

Thaw the chicken livers. Prepare as described on p. 78. Leave whole.
Cut the goose, turkey or duck livers into thick slices. Peel and seed the
grapes.

Heat 2 oz (50 g) of the butter with the oil in a heavy frying pan and
quickly fry the slices of bread on both sides. Set aside and keep warm on
a heated serving dish.

Rinse the pan or wipe with kitchen paper, add the remaining butter
and when it begins to foam put in the livers, spreading them out so that
they cook evenly. Cook over a moderately high heat for no more than 3
minutes on each side – they should be brown on the outside and pink but
set inside. Season generously and spoon on to the fried bread. Keep hot.

Add the wine or vermouth to the pan, stir well to loosen any sediment
and bring quickly to the boil. Boil for 1 minute to reduce. Add the grapes
to the pan and allow to heat through, then pour over the livers. Serve hot.

Warm Salad with Chicken Livers

Serves 4

A delicious and unusual first course which looks particularly pretty if
you use the Italian red radiccio lettuce.

4 oz (100 g) frozen chicken livers	1 tbls (15 ml) wine vinegar
1 lettuce	1 egg yolk
2 oz (50 g) butter	3 tbls (45 ml) oil
1 tsp (5 ml) Dijon mustard	salt and freshly ground pepper

Thaw the chicken livers and prepare as described on p. 78. Leave whole.

Wash and dry the lettuce leaves and arrange them in a salad bowl.

Heat the butter in a frying pan, add the chicken livers and fry for 3 to 4
minutes – they should be just pink and set. Remove from the heat and
chop finely. Return to the pan to keep warm.

Put the mustard into a bowl and stir in the vinegar. Add the egg yolk
and stir until smooth. Dribble in the oil, stirring all the time to combine
well and make a smooth, thick dressing. Season to taste, then stir in the
chicken livers.

Pour over the lettuce and serve. Toss the salad at the table.

Chicken Liver Timbales with Mushroom Sauce

Serves 6 as a starter

This is a good starter for a dinner party, as the mixture can be made in advance, and kept in a covered bowl in the refrigerator for an hour or so until you are ready to cook it.

8 oz (225 g) frozen chicken livers
1 oz (25 g) butter
1 oz (25 g) plain flour
4 fl oz (125 ml) milk
salt and freshly ground pepper
1 whole egg
1 egg yolk
3 tbls (45 ml) single or
    whipping cream

1 tbls (15 ml) madeira or port

Mushroom sauce
8 oz (225 g) mushrooms
1 oz (25 g) butter
2 tsp (10 ml) cornflour
¼ pt (150 ml) single or
    whipping cream
salt and freshly ground pepper

Thaw the chicken livers and prepare as described on p. 78.

Heat the oven to 180°C, 350°F, gas 4.

Melt the butter in a small saucepan, add the flour and stir over a gentle heat for 2 to 3 minutes without allowing the mixture to brown. Stir in the milk, season and continue to stir until the sauce thickens. Remove from the heat and allow to cool.

Put the chicken livers into a blender or food processor with the whole egg and egg yolk, a little salt and freshly ground pepper. Blend at maximum speed for about 1 minute, until the mixture is absolutely smooth. Add the cream, the madeira or port and the béchamel sauce. Blend again and check the seasoning.

Divide the mixture between 6 well-buttered ramekins, place in a roasting tin and pour in boiling water to come halfway up the sides of the ramekins. Bake in the oven for about 30 minutes until the timbales have risen well, are slightly golden round the sides and a skewer inserted into the middle comes out clean.

While the timbales are cooking, make the mushroom sauce. Put the finely sliced mushrooms in a small saucepan with the butter. Cook gently for about 5 minutes until soft. Sprinkle in the cornflour and cook, stirring for a further 3 to 4 minutes. Add the cream, season, and stir over a low heat until the sauce is thick and creamy.

Turn the timbales out on to individual plates and spoon mushroom sauce over each one.

Chicken Liver Risotto

Serves 4

White rice should be used for this recipe, as natural brown rice would take too long to cook.

8 chicken livers
2 oz (50 g) butter
1 medium onion
3–4 oz (75–100 g) mushrooms
14 oz (400 g) long-grain or
  Italian risotto rice

about 2 pts (1.2 L) chicken stock
salt and freshly ground pepper
2 tbls (30 ml) chopped parsley

grated parmesan cheese

Allow frozen chicken livers to thaw and cut each into eight pieces.

Melt the butter in a large heavy frying pan and fry the livers briskly until they stiffen. Remove from the pan and keep warm.

Add the finely chopped onion to the pan and cook gently until transparent. Add the finely chopped mushrooms and cook for a further 2 minutes or so. Add the rice and stir over a gentle heat for no more than 2 minutes, making sure it does not stick to the pan.

Gradually stir in the boiling stock a cupful at a time, keeping the pan over a steady heat and allowing each cup of stock to be absorbed by the rice before you add the next. You may need rather more or less stock – the rice should be cooked until tender but not mushy.

Season, mix in the chicken livers and the parsley and serve on a very hot dish, handing round the grated cheese.

KIDNEYS

Kidneys must be allowed to thaw before cooking, as otherwise they become very tough, but they may be prepared for cooking as soon as they have thawed sufficiently. A 1-lb (500-g) pack will take approximately 3 hours at room temperature, or overnight in the refrigerator.

Skin the kidneys (their encasing of suet fat has usually already been removed, but they are generally still inside a very thin membrane which will peel off quite easily). Halve or slice them while they are still partially frozen, and cut out the central fatty core.

Kidneys in Wine and Cream Sauce

Serves 4

This simple dish is excellent served with rice and a green salad.

1 lb (500 g) frozen lamb's kidneys	¼ pt (150 ml) white wine
6–8 shallots or 2 medium onions	1 tsp (5 ml) French mustard
8 oz (225 g) mushrooms	salt and freshly ground pepper
4 oz (100 g) butter	2 tbls (30 ml) double or sour cream
1 tbls (15 ml) plain flour	squeeze of lemon juice

Part-thaw the kidneys and skin and core them (see p. 81). Slice thickly.

Finely chop the shallots or onions, and thinly slice the mushrooms.

Melt half the butter in a heavy frying pan, add the kidneys and cook until they change colour. Remove from the pan and keep warm.

Add the remaining butter to the frying pan and cook the shallots until soft and golden. Add the mushrooms, raise the heat and cook quickly for 2 minutes. Remove from the pan and keep warm.

Sprinkle the flour into the frying pan, stir and cook without browning for 2 minutes. Add the wine, stir well to make a smooth sauce, and incorporate any sediment on the bottom of the pan. Return the kidneys to the pan, cover and simmer for 5 minutes or until they are cooked.

Stir in the mushroom and shallot mixture, add mustard and salt and pepper to taste, then add the cream and lemon juice. Bring quickly to the boil and serve very hot.

Kidneys Flambés

Serves 3–6

Serve as a starter at the beginning of the meal, as a savoury at the end, or as a simple and quick, but marvellously good, supper dish for 3 or 4 people.

1 lb (500 g) frozen lamb's kidneys	pinch of cayenne
8 shallots	pinch of dried thyme
6 oz (175 g) butter	salt and freshly ground pepper
1 tbls (15 ml) oil	¼ pt (150 ml) dry white wine
3–4 slices of bread	
2 tbls (30 ml) brandy	*1 tbls (15 ml) finely chopped parsley*

Part-thaw the kidneys and skin and core them (see p. 81). Slice thinly.

Chop the shallots finely.

Heat half the butter with the oil in a heavy frying pan and quickly fry the pieces of bread until crisp and golden on both sides. Cut into triangles and keep warm.

Rinse the frying pan and melt 2 oz (50 g) of the remaining butter, add the shallots and fry until softened. Add the kidneys and cook over a fairly high heat for 2 minutes, stirring constantly.

When the kidneys have just changed colour, add the brandy and set alight. Remove from the heat and wait for the flames to die down.

Add the cayenne, thyme, salt and pepper. Return to a moderate heat and cook for a further 2 minutes, until the kidneys are just cooked. Add the wine, stir well to amalgamate, bring to the boil and let the sauce bubble for a minute. Add the remaining butter, cut into small pieces, to give the sauce a glaze.

Pour over the fried bread triangles, sprinkle with the parsley and serve very hot.

Devilled Kidneys

Serves 4

For supper, or for a special breakfast or brunch.

8 frozen lamb's kidneys	pinch of cayenne
2 oz (50 g) butter	good squeeze of lemon juice
2 tsp (10 ml) made English mustard	4 slices bread
2 tsp (10 ml) French Dijon mustard	
¼ tsp (1.2 ml) chilli powder	

Part-thaw the kidneys. Skin and split them lengthways from the smooth side, not quite cutting through, so that you can open them out into a circle, and carefully remove the fatty core from the centre. Skewer each kidney with two thin skewers or wooden cocktail sticks, so that they can be kept flat on the grill pan grid.

Mash the butter to a smooth paste with the remaining ingredients, except the bread, and spread a little of the mixture on to each kidney.

Toast the bread and keep warm.

Just before you are ready to serve, heat the grill to almost its fiercest heat and grill the kidneys for about 4 minutes on each side, dotting with a little more seasoned butter on the second side.

Spread the remaining butter mixture on the pieces of toast, heat them through briefly if necessary and place two kidneys, sizzling hot, on each slice.

Serve at once.

Kidney and Bacon Rolls

Serves 4

These rolls can either be made quite small, for serving with drinks, or larger, for a main course. In the former case you will need about 16 narrow rashers of bacon, in the latter 8 wide rashers. The bacon should be lean and very thinly sliced.

8 frozen lamb's kidneys	1 egg
2 medium onions	3 tbls (45 ml) chopped parsley
½ oz (15 g) butter	squeeze of lemon juice
1 tbls (15 ml) oil	salt and freshly ground pepper
2½ oz (65 g) fresh breadcrumbs	12 oz (350 g) bacon rashers

Part-thaw the kidneys, skin, cut in half lengthways and remove the fatty core.

Heat the oven to 190°C, 375°F, gas 5.

Cook the finely chopped onions gently in the butter and oil until soft and golden. Mix with the breadcrumbs, beaten egg, parsley and lemon juice and season to taste. Spread this mixture on the bacon rashers.

Put either a whole or half a kidney on each rasher of bacon, according to whether you are using wide or narrow rashers, roll up and secure with a wooden cocktail stick. Cook in the oven for 20 to 25 minutes. Serve very hot.

HEART

Braised Heart

Serves 4–6

Pig's, calf's or lamb's hearts can all be used for this economical but really tasty dish, known in Denmark as 'Passionate Love'.

2 lb (1 kg) frozen hearts	salt and freshly ground pepper
4 oz (100 g) fresh breadcrumbs	1 oz (25 g) plain flour
1 small onion	1 tsp (5 ml) wine vinegar
1 tbls (15 ml) finely chopped parsley	1 tsp (5 ml) French mustard
finely grated rind of 1 lemon	1 pt (600 ml) stock
1 egg	1 bay leaf
2 tbls (30 ml) milk	¼ pt (150 ml) double or sour cream
2 oz (50 g) butter	

Allow the hearts to thaw and remove any fat, arteries and veins.

Mix the breadcrumbs with the finely chopped onion, parsley and lemon rind, and bind together with the egg lightly beaten in the milk and half the butter, melted. Season.

Stuff the heart cavities with this mixture and secure the openings with skewers or wooden cocktail sticks. Roll the hearts in the well-seasoned flour.

Melt the remaining butter in a flameproof casserole, sprinkle in any remaining flour, allow to brown a little, stirring, add the vinegar and mustard, then slowly stir in the stock to make a smooth sauce.

Add the hearts and the bay leaf, cover and simmer over a very low heat, or in a moderate oven (180°C, 350°F, gas 4), for 2 to 3 hours, until the hearts are quite tender.

Stir in the cream, check the seasoning and serve hot with noodles or creamed potatoes.

Chicken

✳✳✳✳✳✳✳✳✳✳✳✳✳✳✳✳✳✳✳

Chickens are a splendid stand-by to have in the freezer, especially as there has recently been a great advance in the rearing and selling of corn-fed birds, which have a much better taste and texture than the rubbery chickens which were all that most of us were previously able to buy. Most of these corn-fed birds come from abroad; at the present moment only one British firm, Moy Park, rears them here, in three different varieties – corn-fed, *poulet noir* (which tastes deliciously like guinea-fowl) and free range. They are available from most supermarkets, and cost a little more, but are worth it, not least because the birds are kept in more humane conditions than the run-of-the-mill oven-ready chicken.

Whatever chickens you buy, it is best to get them fresh and freeze them yourself (the corn-fed chickens are mostly sold fresh in any case). Unless you have a very large freezer, or want to cook the chickens whole, it is more economical of space to joint them before freezing. In this way, too, they thaw much more quickly. Or you can buy fresh joints for freezing – legs, wings and breasts – and these too are available from corn-fed birds.

All frozen poultry must be completely thawed before cooking. And never stuff birds before freezing, as the thawing process then takes so long that there is time for any food-poisoning organism to develop.

If possible, see that the giblets are included when you buy whole birds. The livers can be prepared and frozen separately for future use (see p. 78), and the heart and gizzard are excellent for gravy and stock.

Chicken Casserole Provençale

Serves 4

This is a very simple and economical way of cooking chicken joints. It is also excellent made with rabbit.

4 chicken joints	1 sprig of rosemary
1 × 14-oz (400-g) can tomatoes	salt and freshly ground pepper
4 oz (100 g) black olives	

Frozen chicken joints must be allowed to thaw.

Put the chicken joints into a flameproof casserole. Pour over the tomatoes with their juice. Add the stoned olives together with the rosemary. Season.

Bring slowly just to the boil, cover the casserole and cook over gentle heat, barely simmering, for 1½ hours.

Chicken with Pimientos and Rosemary

Serves 6

1 × 4½-lb (2-kg) chicken or chicken joints	1 medium green pimiento
1 tbls (15 ml) seasoned flour	4–5 sprigs of rosemary
2 oz (50 g) butter or chicken fat	¼ pt (150 ml) dry white wine
3 tbls (45 ml) olive oil	1 × 14-oz (400-g) can tomatoes
1 medium onion	salt and freshly ground pepper
1 medium red pimiento	2 oz (50 g) cooked ham

Frozen chicken or chicken joints must be allowed to thaw thoroughly.

Joint the chicken if you are using a whole one. Roll the joints in the well-seasoned flour. Heat the butter or chicken fat and 1 tbls (15 ml) of the olive oil in a frying pan and gently fry the chicken joints until they are golden brown. Remove to a flameproof casserole.

Fry the chopped onion for a few minutes until softened.

Add the seeded and fairly finely sliced pimientos to the pan, together with the remaining oil and the rosemary. Fry for about 10 minutes, then add the wine. Simmer for a few minutes to let the wine partly evaporate, then add the tomatoes with their juice. Season lightly and heat through.

Lay the ham, cut in strips, on top of the chicken and pour over it the contents of the frying pan. Cover the casserole and simmer gently for about 45 minutes, until the chicken is tender. Remove the rosemary.

To serve immediately: serve with rice.

To freeze: allow to cool, then freeze in a plastic container.

To serve after freezing: allow to thaw, preferably in the refrigerator for 24 hours, then heat through very gently. Serve as above.

Note: do not freeze this dish if you have used frozen chicken.

Portuguese Chicken

Serves 6

1 × 4½-lb (2-kg) chicken or chicken joints
3 small onions
12 oz (375 g) mushrooms
3 oz (75 g) butter
6 tbls (90 ml) dry white wine or vermouth

scant ½ pt (250 ml) stock
5 tbls (75 ml) tomato purée
1 tsp (5 ml) dried oregano
1 clove garlic
salt and freshly ground pepper

Frozen chicken joints must be allowed to thaw thoroughly.

If you are using a whole chicken, cut it into small pieces (you should be able to finish up with 12 or 13 pieces, cutting the legs, wings and breasts each into two pieces, and perhaps making the wish-bone an extra joint).

Chop the onions finely and slice the mushrooms.

Sauté the chicken in the butter in a flameproof casserole until light brown. Remove and set aside. Add the onions to the casserole and cook gently until tender. Add the chicken, the mushrooms, wine or vermouth, stock, tomato purée, oregano and crushed garlic. Season. Cover, bring gently to the boil and simmer gently for 30 to 45 minutes, until the chicken is tender.

To serve immediately: this dish is quite rich, so serve with plain boiled rice and a green salad.

To freeze: allow to cool, then freeze in a plastic container.

To serve after freezing: allow to thaw in the refrigerator for at least 24 hours. It may take longer, or you may need to leave it at room temperature for 3 to 4 hours to finish the thawing process. Heat very gently until it is warmed through, and serve as above.

Note: do not freeze this dish if you have used frozen chicken.

Chicken Breasts Adriana

Serves 6

A rich and appetizing dish, good for dinner parties and buffet suppers as it will wait happily in the oven until you are ready to serve it.

6 chicken breasts	1 tbls (15 ml) flour
1 egg	a little oil and butter for frying
2 oz (50 g) grated cheese	4 oz (100 g) thinly sliced cooked ham
salt and freshly ground pepper	2 × 6-oz (175-g) mozzarella cheeses

Allow frozen chicken breasts to thaw and cut away the bone and skin. Cut each one in half.

Beat the egg and stir in the grated cheese and a little salt and pepper. Marinate the chicken breasts in this mixture for 3 to 4 hours.

Heat the oven to 100°C, 200°F, gas ¼.

Scrape off the marinade from the chicken breasts and pat them dry. Dust with flour and fry in the butter and oil – they will take only a few minutes on each side.

Arrange the chicken breasts in a single layer in a shallow ovenproof dish. Cover with the ham, then with slices of mozzarella cheese about ¼ inch (5 mm) thick. Heat through in the oven for about 40 minutes.

Serve with mashed potatoes, into which you have beaten the egg and cheese marinade.

Variation: blade bone steak, an extremely economical cut, can be substituted for the chicken breasts. It should be cut very thin, and should be beaten a little before you marinate it. It will need to be fried for only a minute or so on each side.

Chicken Breasts with Mushrooms and Cream

Serves 6–8

Another good dish for a dinner party.

6–8 chicken breasts
2 medium onions
2 oz (50 g) butter or chicken fat
⅓ pt (200 ml) dry white wine
1 lb (500 g) mushrooms

salt and freshly ground pepper
squeeze of lemon juice
⅓ pt (200 ml) double or
  whipping cream

Frozen chicken breasts must be allowed to thaw.

Carefully remove the bones and skin. Cut the breasts in half if they are very large.

Heat the oven to 140°C, 275°F, gas 1.

Chop the onions finely and cook gently in half the fat in a large flameproof casserole until soft and golden. Remove from the pan and set aside. Heat the remaining fat in the pan and brown the chicken breasts on both sides. Remove from the pan and set aside. Add the wine and cook briskly until it has reduced by about half.

Return the chicken breasts and onions to the pan, add the sliced mushrooms, season and add the lemon juice. Stir in the cream and heat slowly until the sauce has almost come to the simmer, but do not let it boil. Cover and cook in the oven for 2 hours.

Serve with plain boiled rice.

Note: a lovely way of serving this dish is to skin 5 or 6 tomatoes, scoop out the middle and leave them upside down to drain. When you are ready to serve the chicken breasts, make a ring of rice round the edge of a big dish and pile the chicken breasts and their sauce in the middle. Spoon a little rice into each tomato, decorate with a small sprig of watercress or parsley, and arrange on top of the rice.

Malayan Chicken

Serves 6

The combination of black olives and water chestnuts makes this an interesting dish, and it looks especially pretty served in a ring of rice.

1 × 4½-lb (2-kg) chicken or chicken joints	1 tsp (5 ml) ground ginger
	pinch of garlic salt
8 oz (225 g) black olives	scant ½ pt (250 ml) dry white wine
1 × 8-oz (225-g) can water chestnuts	salt and freshly ground pepper
2–3 tbls (30–45 ml) oil	1 tbls (15 ml) water
⅓ pt (200 ml) orange juice, fresh or canned	1 tbls (15 ml) cornflour

a few sprigs of parsley

Frozen chicken or chicken joints must be allowed to thaw thoroughly.

Stone the olives and chop roughly. Drain the water chestnuts and cut in half.

If you are using a whole chicken, cut into joints. Brown the chicken in the oil in a flameproof casserole. Reduce the heat and add the orange juice, ginger, garlic salt, wine, olives and water chestnuts. Season, cover and simmer for 30 to 45 minutes, until the chicken is tender.

Mix together the water and the cornflour to make a smooth paste and gently stir into the casserole until the sauce thickens.

To serve immediately: serve with rice, and garnish with sprigs of parsley to give colour.

To freeze: allow to cool. Freeze in the casserole if you can spare it, but if not, in a plastic container.

To serve after freezing: leave in the refrigerator for at least 24 hours. It may take longer to thaw, or you may need to leave it at room temperature for 3 to 4 hours to finish the thawing process, since it should be completely thawed before reheating. Heat very gently until it is warmed through. Serve as above.

Note: do not freeze this dish if you have used frozen chicken.

Tandoori Chicken

Serves 6

An excellent way to use chicken joints. Best made with strained or Greek yoghurt, which is thicker than the usual bought variety. A ready blended mixture of the spices used for tandoori cooking is available from most delicatessen or Indian stores.

6 chicken breasts or	½ pt (300 ml) yoghurt
12 chicken thighs	½–1 tbls (7.5–15 ml) tandoori
juice of ½ lemon	spice mixture
salt	1 tbls (15 ml) oil

Allow frozen chicken pieces to thaw sufficiently to be able to skin them. With a sharp, pointed knife, make shallow incisions along the top of each piece. Arrange in a single layer in a shallow ovenproof dish, and sprinkle generously with lemon juice and salt.

In a separate bowl mix the yoghurt with the spice mixture and oil, and smear this liberally over the chicken pieces. Cover with clingfilm and leave overnight in the refrigerator to allow the chicken to thaw and become impregnated with the spices.

Heat the oven to 190°C, 375°F, gas 5 and bake the chicken for 40 to 50 minutes. Serve hot, with rice and a green salad.

Chicken Fricassée

Serves 6–8

This is a useful dish for entertaining, especially if you have some cooked chicken in the freezer. It looks pretty, and can be made in large quantities almost as quickly as in small ones.

1 lb (500 g) cooked chicken	salt and freshly ground pepper
1 lb (500 g) mushrooms	1 × 6½-oz (190-g) can pimientos
4 oz (100 g) butter	lemon juice
2 oz (50 g) plain flour	
1 pt (600 ml) chicken or veal stock	

If the cooked chicken has been frozen, allow to thaw, and cut into largish bite-sized pieces.

Slice the mushrooms about ⅛ inch (2.5 mm) thick. Put them into a frying pan with half the butter, cover, and cook gently until they are soft.

Melt the remaining butter in a flameproof casserole, add the flour, and cook gently for 2 to 3 minutes, stirring, without allowing to brown. Add the stock and bring to the boil, stirring, until the sauce is smooth and thick. Season with salt and plenty of pepper, and add the chicken, mushrooms and pimientos, drained, and cut into strips about ¼ inch (5 mm) wide. Be careful to remove any seeds that may have been left in them. Add lemon juice to taste – you will probably need plenty, to counteract the sweet taste of the chicken.

Warm through over a low heat until bubbling.

To serve immediately: serve with rice or mashed potatoes and a green salad.

To freeze: allow to cool, then freeze in a plastic container.

To serve after freezing: allow to thaw, then reheat as slowly as possible. Serve as above.

Note: do not freeze this dish if the cooked chicken has already been frozen.

Moroccan Chicken

Serves 6

1 × 4½-lb (2-kg) chicken or chicken joints	½ tsp (2.5 ml) ground turmeric
2 oz (50 g) butter	¼ tsp (1.2 ml) ground coriander
1 tbls (15 ml) olive oil	2 tomatoes, fresh or canned
2 onions	½ pt (300 ml) water
2 cloves garlic	salt
½ tsp (2.5 ml) freshly ground black pepper	1 lemon, preferably a preserved lemon*
1 tsp (5 ml) ground ginger	4 oz (100 g) green olives

Frozen chicken must be allowed to thaw. If you are using a whole chicken, cut into joints.

Heat the butter and oil in a flameproof casserole and brown the chicken joints on all sides. Remove with a slotted spoon and set aside.

Add the finely chopped onions and garlic to the casserole, together with the pepper, ginger, turmeric and coriander. Cook for 2 minutes, stirring to blend, then return the chicken pieces and turn them well.

Skin and quarter the fresh tomatoes, or chop canned ones roughly,

---

* Lemons preserved in salt can be bought in Greek and some delicatessen shops. To make your own, wash 5 or 6 lemons well, prick all over quite deeply with a skewer, and pack tightly into a jam or preserving jar. Add 5–6 tbls (75–90 ml) coarse cooking salt, top up the jar with the juice of 3 to 4 more lemons, cover, shake well, and leave to mature for 2 to 3 weeks. Rinse each lemon before using.

These preserved lemons are delicious with grilled meats or fish, or added, finely chopped, to salads. Or add a few pieces to olives when serving them with drinks.

add to the casserole and raise the heat to evaporate the liquid. Add just under half the water and boil rapidly, stirring well to scrape any sediment off the bottom of the casserole. Over a moderate heat slowly add the remaining water, stirring constantly, until the sauce thickens. When all the water has been added, season with a little salt, cover and simmer gently until the chicken is tender.

Grate the rind of the lemon and scoop out the flesh, or rinse the preserved lemon and cut into small pieces. Stone the olives if necessary. Add the lemon and olives to the casserole 5 minutes before serving and heat through.

If the dish seems too liquid, raise the heat and cook uncovered for a few minutes to reduce. Check the seasoning before serving.

Spicy Chicken Wings

An unusual combination of spices makes this an interesting, as well as economical and quickly made, family supper dish.

Allow 2–3 chicken wings per person.

chicken wings	powdered ginger
salt	paprika
freshly ground black pepper	oil

Frozen chicken wings must be allowed to thaw completely.

Sprinkle each chicken wing with salt, pepper, ginger and paprika, and arrange in the grill pan. Dribble a little oil over each.

Place under a hot grill, and when the skin begins to crisp, turn down the heat and continue to grill for a further 15 to 20 minutes, turning over once, until the wings are well cooked. Turn up the heat again for a few minutes at the end to brown the tops before serving.

Cooking for Large Numbers

✳✳✳✳✳✳✳✳✳✳✳✳✳✳✳✳✳✳✳

The freezer is one of the cook's greatest friends when it comes to catering for large numbers. Perhaps the school holidays are looming; or you are abandoning the family for a while and don't want to leave them foodless; or maybe you have a party in mind – whatever the reason, it is wonderful to be able to plan and shop ahead and know that you can cook at your leisure, freeze, and have the dishes ready when they are needed.

The recipes which follow are all intended for occasions such as these, although of course the quantities can be reduced for smaller numbers. Since the idea is that the dishes should be cooked in advance and stored in the freezer, fresh, not frozen, meat has been specified in every case.

Lasagne al Forno

Serves 12–18

This deliciously creamy mixture of pasta and meat with sauce, though time-consuming to prepare, is easy to serve to large numbers of guests, and is an ideal dish for a buffet party.

1 lb (500 g) lasagne sheets (green ones provide an attractive colour)
1 tbls (15 ml) oil

Meat sauce
8 oz (225 g) bacon or smoked ham, diced
2 onions, finely chopped
2 carrots, finely chopped
2 celery stalks, finely chopped
2 cloves garlic, finely chopped

1 glass red or white wine
3 lb (1.5 kg) minced meat (preferably veal, but veal and pork, or even beef, will do)
½ pt (300 ml) stock
1 tbls (15 ml) tomato purée
salt and freshly ground pepper
pinch of freshly grated nutmeg
8 oz (225 g) chicken livers (optional)

Cheese sauce
8 oz (225 g) butter or margarine
6 oz (175 g) plain flour
3¾ pts (2 L) milk
salt and freshly ground pepper
pinch of freshly grated nutmeg
1 bay leaf
1 lb (500 g) grated parmesan,
    cheddar or other hard cheese

Tomato sauce
1 lb (500 g) onions, finely chopped
2 large cloves garlic, finely chopped
1 oz (25 g) butter
1 tbls (15 ml) oil
3 × 5-oz (150-g) cans tomato purée
1 glass red wine
salt and freshly ground pepper
1 cube sugar

To make the meat sauce, sweat the bacon or ham in a large heavy frying pan with a lid. Add the onions, carrots, celery and garlic and fry gently. Pour on the wine and let it bubble for a minute before adding the minced meat, stock, tomato purée, seasoning and nutmeg. Cover and leave to simmer.

If you are using chicken livers fry them separately in a little butter and chop quite finely. Add to the meat sauce before assembling the dish.

To make the cheese sauce, melt the butter or margarine in a saucepan, add the flour and cook, stirring, for 2 to 3 minutes, without allowing to brown. Slowly stir in the milk, season, add the nutmeg and bay leaf and leave to thicken at the back of the stove. Remove the bay leaf and stir in half the grated cheese just before using the sauce.

To make the tomato sauce, soften the onions and garlic in the butter and oil in a saucepan, then add the tomato purée, wine, seasoning and sugar. Leave to simmer until thickened.

Now cook the lasagne sheets for 10 to 12 minutes, according to the instructions on the packet, in fast-boiling, salted water to which you have added 1 tbls (15 ml) oil. Put the sheets singly into the water and cook in batches, as they must move freely. (If you are using fresh lasagne, it will take less time to cook.) When the lasagne is cooked, drain and rinse in cold water, and spread the sheets out on clean tea cloths to dry.

Use well-greased, ovenproof china, earthenware or foil dishes, 2½–3 in (6–8 cm) deep. Add the chicken livers, if you are using them, to the meat sauce and stir in about ½ pt (300 ml) of the cheese sauce. Then, starting with the lasagne, make alternate layers of pasta and meat sauce, leaving a good 1-in (2.5-cm) space at the top of each dish. Spread on a layer of tomato sauce and finish with a thick layer of the cheese sauce. Sprinkle the remaining cheese over the top.

To serve immediately: cook in a moderately hot oven (190°C, 375°F, gas 5) for 1 to 1½ hours, until heated through and crisp on top. Serve at once.

To freeze: leave the dishes to cool, then wrap in foil and freeze.

To serve after freezing: if possible allow to defrost overnight in the refrigerator, or for 3 to 4 hours at room temperature. Then cook as for serving immediately. You can, if necessary, cook the lasagne straight from the freezer, but allow about 1 hour extra cooking time.

Jo Mazotti

Serves 12–18

The perfect dish for buffet parties, especially for occasions such as New Year's Day when something unusual but 'comforting' is required. It is very easy to make in really large quantities, and can be placed straight from the freezer into the oven, so no last-minute preparation is required. Serve with salad.

2 lb (1 kg) onions
1 carrot
1 celery stalk
2 large green pimientos
1 lb (500 g) mushrooms
8 oz (225 g) butter
2 tbls (30 ml) oil
2 lb (1 kg) minced pork or
    pork and beef
2 × 5-oz (150-g) cans concentrated
    tomato purée

juice of 1 lemon
salt and freshly ground pepper
1 lb (500 g) short-cut macaroni or
    noodle twists
1½ lb (750 g) sharp cheese such as
    mature farmhouse or
    Canadian cheddar
4 oz (100 g) fine dried breadcrumbs

Slice the onions. Chop the carrot and celery finely. Seed the pimientos and dice quite finely. Slice the mushrooms. Heat half the butter with the oil in a heavy flameproof casserole or saucepan. Add the onions, carrot and celery and fry until golden-brown.

Add the meat and cook over a fairly high heat, stirring frequently, until all the meat has changed colour. Add the sliced mushrooms, the seeded and diced peppers, the tomato purée and the lemon juice and season to taste. Simmer for 10 to 15 minutes until well amalgamated.

Meanwhile cook the macaroni or noodles in plenty of boiling, salted water until just tender. Be sure not to overcook, and cook in several batches if necessary; the pasta should be able to move freely in the water so that it cooks evenly and does not stick together. As soon as it is ready, drain and rinse in cold water. Leave to drain.

Cut 1 lb (500 g) of the cheese into cubes the size of small sugar cubes. Grate the remainder and blend with the breadcrumbs.

Mix the pasta and the cubed cheese with the meat mixture, check for seasoning and pour into shallow ovenproof dishes. Sprinkle with the grated cheese and breadcrumbs.

To serve immediately: dot with the remaining butter, and put into a hot oven (220°C, 425°F, gas 7) for 15 minutes. Turn the oven down to 190°C, 375°F, gas 5, and cook for at least 1 further hour. It is virtually impossible to overcook this dish, so timing does not have to be too accurate.

To freeze: allow to cool, then wrap in foil and freeze.

To serve after freezing: you can either allow the dish to thaw for a while at room temperature, or you can transfer it straight from the freezer into the oven. Cook as above, but allow 30 minutes longer in the hot oven if the dish has not been allowed to thaw.

Lamb and Leek Stew

Serves 10–12

A good casserole for a winter's day. A shoulder of lamb can be used for this dish. Get your butcher to bone it and give you the bones. You will need about 4–5 oz (100–150 g) lean meat per person.

2 large onions	4 tbls (60 ml) oil
2 lb (1 kg) leeks, trimmed weight	1 pt (600 ml) chicken or veal stock
6 ripe tomatoes	salt and freshly ground pepper
3 lb (1.5 kg) lean lamb, trimmed weight	

Slice the onions. Discard most of the green part of the leeks, cut into 1-in (2.5-cm) pieces, and wash very well so that any grit is removed. Skin and seed the tomatoes and chop them roughly.

Trim the fat off the meat and cut into bite-sized pieces. Heat the oil in a flameproof casserole and sauté the lamb, stirring until lightly browned.

Add the vegetables to the casserole and stir over a gentle heat for 5 minutes. Add the hot stock and season. Put bones on top to give flavour. Bring to the boil, cover and simmer very gently for 45 minutes, or until the meat and vegetables are tender. It may be necessary to skim the stew when it first comes to the boil.

To serve immediately: remove the bones before serving the stew with mashed potatoes.

To freeze: allow to cool, remove the bones, then freeze in a plastic container.

To serve after freezing: remove the stew from the freezer 24 hours before you want to serve it, and leave in the refrigerator. It should have completely thawed before you reheat it, so if necessary take out of the refrigerator and leave at room temperature for 2 to 3 hours before reheating. Bring slowly to the boil and serve as soon as it is heated through.

Chilli con Carne

Serves 10–12

An excellent, sustaining and inexpensive dish, ideal for a cold evening. Serve with plain boiled rice or noodles and a green salad.

1 lb (500 g) red kidney beans	2 tsp (10 ml) chilli powder
2 tbls (30 ml) oil	2 lb (1 kg) minced beef
2 onions	1 lb (500 g) tomatoes, fresh or
2 cloves garlic	canned
2 tsp (10 ml) ground turmeric	salt and freshly ground pepper
1 tbls (15 ml) ground coriander	

Soak the beans in cold water for 4 hours or overnight.

Heat the oil in a flameproof casserole and gently fry the chopped onions and garlic until golden-yellow. Add the spices and fry for a further 5 minutes.

Add the meat, raise the heat and stir well until all the meat has changed colour. Add the skinned and roughly chopped tomatoes with their juice, season, then cover and simmer for about 1 hour.

Meanwhile drain the beans, put into a large saucepan and cover with fresh cold water. Bring slowly to the boil, cover and simmer for 1 hour or until tender. Drain and mix with the meat.

To serve immediately: check the seasoning and allow to simmer together for 15 minutes before serving.

To freeze: cool, then freeze in the casserole or in convenient containers.

To serve after freezing: allow to thaw for a few hours at room temperature or preferably overnight. Reheat, gently at first, then bring slowly to the boil and boil for a minute or two. Check the seasoning and simmer for 30 minutes before serving.

Dhal

Serves 10–12

Dhal, a mess of pottage, or more properly of lentils, seems to be the Indian equivalent of chips – it comes with everything. It is delicious on its own, or with plain boiled rice, or as a vegetable side dish to serve with a curry. It freezes extremely well.

1 lb (500 g) lentils	salt
2 pts (1.2 L) water	2 tomatoes
2 tbls (30 ml) vegetable oil	juice of ½ lemon or 1 lime
2 onions	2 tbls (30 ml) water
½ oz (15 g) fresh ginger root	
3 cloves garlic	*2 tbls (30 ml) oil*
2 tsp (10 ml) ground coriander	*2 onions*
4 tsp (20 ml) ground turmeric	*1 tsp (5 ml) whole cumin seeds*

Wash the lentils thoroughly. Bring them to the boil in the water, skim, cover and simmer until most of the water has been absorbed. This may take as long as 1 hour, depending on how fresh the lentils are.

In a large frying pan heat the oil and soften the roughly chopped onions. Add the finely chopped ginger and garlic, the spices and salt, the skinned and roughly chopped tomatoes, the lemon or lime juice and the 2 tbls (30 ml) water.

Cook over a gentle heat for 5 minutes to amalgamate, then stir into the lentils, mix well and continue to cook for 5 to 10 minutes, until the flavours have been absorbed and all the liquid has evaporated. The consistency should be that of a very thick soup.

To serve immediately: heat the oil in the frying pan and cook the finely sliced onions over a high heat until they are brown and crisp. Add the cumin seeds at the last moment and let them brown and pop. Pour over the lentil mixture just before serving.

To freeze: cool and freeze in convenient quantities in waxed or plastic containers.

To serve after freezing: heat the dhal through gently from frozen, adding a little more water if necessary to stop it from catching. Serve as above.

Moussaka

Serves 12

This is a particularly light moussaka, with a very fresh taste of aubergines and tomatoes.

3 lb (1.5 kg) aubergines
salt
1 lb (500 g) onions
4 oz (100 g) butter or chicken fat
2 tbls (30 ml) plain flour
2 lb (1 kg) minced beef
½ pt (300 ml) red wine
  (or stock or water)
4 tbls (60 ml) tomato purée

4 oz (100 g) sultanas (optional)
2 tsp (10 ml) sugar
2 tbls (30 ml) chopped parsley
¾–1 pt (450–600 ml) olive oil or
  half olive oil, half vegetable oil
1 × 14-oz (400-g) can tomatoes
2 tbls (30 ml) chopped fresh
  marjoram or basil
freshly ground pepper

Sauce
2 oz (50 g) butter
2 oz (50 g) plain flour
1½ pts (900 ml) milk
6 oz (175 g) grated cheddar cheese

2 eggs, separated
salt and freshly ground pepper

Cut the aubergines into slices about ¼ in (5 mm) thick. Place in a colander, sprinkle lightly with salt, put a weight on top and leave for about 30 minutes to drain.

Meanwhile cook the finely chopped onions gently in the butter or chicken fat in a large saucepan until soft and golden. Stir in the flour and continue to cook over a low heat for a few minutes. Add the meat, stirring it in well until it changes colour. Add the wine, tomato purée, the sultanas, if you are using them, the sugar and parsley. Cover and simmer gently for about 20 minutes.

While the meat is cooking pat the aubergines dry and fry in batches in the oil in a large non-stick frying pan until golden, adding more oil if necessary. Drain the aubergines on kitchen paper.

Stir the canned tomatoes with their juice and the marjoram or basil into the meat mixture. Season.

Butter one or two casseroles and put in layers of the aubergines and mince, starting and finishing with aubergines.

To serve immediately: cook in a moderately hot oven (190°C, 375°F, gas 5) for about 45 minutes to 1 hour, according to whether you are using one casserole or two, until the moussaka is hot and bubbling. Pour the sauce on top (see below) and return to the oven for no more than 10 to 15 minutes. Serve at once.

To freeze: allow to cool, then wrap the casseroles in foil and freeze.

To serve after freezing: thaw in the refrigerator for at least 24 hours – it may take longer than this, and you may need to keep the moussaka at room temperature for some hours until it has thawed completely. Then proceed as above.

Sauce

Melt the butter in a saucepan, add the flour and stir for 2 to 3 minutes over a low heat. Gradually stir in the milk and bring to the boil. Stir in the grated cheese until melted. When the sauce is smooth, creamy and thick, stir in the egg yolks. Season. Finally fold in the well-beaten whites.

Mexican Spiced Mince

Serves 10–12

Serve in a large bowl placed in the centre of the table; provide a quantity of warmed pitta bread or Mexican tortillas, and let everyone dip in as they please. Followed by a salad and fruit, this is an excellent, economical way to feed a large, informal party.

1 lb (500 g) onions	1 red or green pimiento
3–4 cloves garlic	½ tsp (2.5 ml) ground cinnamon
4 tbls (60 ml) oil	pinch of ground cloves
3 lb (1.5 kg) minced beef	salt and freshly ground pepper
1 lb (500 g) tomatoes	4 oz (100 g) raisins (optional)
2 fresh green or red chillies	

Soften the finely chopped onions and garlic in the oil in a large saucepan. Add the meat, raise the heat and stir until it has all changed colour.

Add the skinned and roughly chopped tomatoes and the seeded and finely chopped chillies and pimiento to the pan. Add the spices and seasoning, cover and cook over a gentle heat for 20 minutes.

Check the seasoning and add the raisins plumped in hot water and drained, if you are using them.

To serve immediately: simmer for a further 10 to 15 minutes and serve as above.

To freeze: cool, then freeze in convenient quantities in wax or plastic containers.

To serve after freezing: allow to thaw for at least 1 to 2 hours, then heat through gently and serve as above. If you have to heat from frozen you may need to add a little water or stock.

Shepherd's Pie

Serves 12

A well-made shepherd's pie needs no apology. It is delicious, bearing no resemblance to the school dinners we all remember with horror.

2 lb (1 kg) cooked lamb or beef, trimmed weight	salt and freshly ground pepper
4 oz (100 g) butter	3 lb (1.5 kg) potatoes
4 medium onions	about ½ pt (300 ml) milk
4 tbls (60 ml) tomato purée	3–4 oz (75–100 g) butter
½ pt (300 ml) dry white wine	
about ¾ pt (450 ml) beef stock	*1 oz (25 g) butter*

Mince the meat. Melt the butter in a large saucepan and gently cook the finely chopped onions until soft and golden. Stir in the meat, tomato purée and wine, and cook for a few minutes. Add enough stock to make the mince quite sloppy, as it will dry a little during the baking stage. Season.

Cook the potatoes in salted water until tender. Drain, return to the pan and leave over a very low heat for 2 to 3 minutes to dry. Mash well. Add enough hot milk and butter to give a creamy consistency, and continue beating vigorously until fluffy. Season generously with pepper.

Put the mince into two or more buttered ovenproof dishes and spoon the potato over the top.

To serve immediately: dot with butter and cook in a moderate oven (180°C, 350°F, gas 4) for 40 minutes to 1 hour, according to whether you have made a small or a large pie, until hot and bubbling and the potato topping is golden-brown.

To freeze: allow to cool, then wrap in foil and freeze.

To serve after freezing: allow to thaw and heat through as above. Or, if time is short, transfer the pie straight from the freezer (having first dotted it with butter) into a moderate oven (180°C, 350°F, gas 4) and cook for at least 1½ hours.

Variation: instead of all potatoes, use a mixture of potatoes and celeriac, swedes, or any other root vegetable.

Potato and Cheese Casserole

Serves 8–10

A good vegetarian supper dish in itself, or excellent as an accompaniment to cold meat or a casserole. The quantities given below will fill 1 × 2-pt (1.2 L) dish or 2 × 1-pt (600-ml) dishes.

2 lb (1 kg) potatoes	1 bay leaf
1 lb (500 g) onions	2 oz (50 g) butter
10 oz (275 g) cheddar or gruyère cheese	salt and freshly ground pepper
½ pt (300 ml) milk or stock	*¼ pt (150 ml) single cream*

Peel and slice the potatoes very finely, using the thinnest blade on a mandolin or the slicing blade of a grater or a food processor. Soak the potato slices in cold water to rinse off some of the starch while you prepare the remaining ingredients.

Cut the onion into thin rings and thinly slice the cheese. Bring the milk or stock to the boil with the bay leaf, then remove from the heat.

Butter liberally a 2-pt (1.2 L) ovenproof dish (or two smaller dishes). Pat the potatoes dry on a clean tea towel and put a layer in the bottom of the dish, followed by a layer of onion slices and a layer of cheese. Dot with a little butter and sprinkle with salt and pepper. Repeat the layers until all the ingredients are used up, finishing with a layer of potatoes topped with some cheese.

Pour over the hot milk or stock (discarding the bay leaf), dot with the remaining butter and cook in a moderately hot oven (190°C, 375°F, gas 5).

To serve immediately: bake until the potatoes are creamily cooked and the top is crisply browned (about 1 hour). Cover with buttered paper towards the end if necessary. Spoon on the cream towards the end of the cooking time.

To freeze: cook for 40 minutes only, then cool wrap in foil and freeze.

To serve after freezing: thaw in the refrigerator overnight, or for 3 to 4 hours at room temperature. Cook as above for 30 to 40 minutes until the casserole has heated through and the top has browned. Spoon on the cream towards the end of the cooking time.

Macaroni Cheese

Serves 8–10

A macaroni cheese destined for the freezer should have more than its usual share of carefully made sauce. The following quantities will fill 1 × 2-pt (1.2 L) dish or 2 × 1-pt (600-ml) dishes.

4 oz (100 g) butter or margarine	8 oz (225 g) short-cut macaroni
4 oz (100 g) plain flour	2 oz (50 g) fresh breadcrumbs,
2 pts (1.2 L) milk	crushed cornflakes or Weetabix
12 oz (350 g) grated cheddar cheese	
salt and freshly ground pepper	*1 oz (25 g) butter*

Melt the butter or margarine in a large, heavy saucepan, stir in the flour and cook over a low heat, stirring, for 2 to 3 minutes. Slowly add the milk, stirring well, until the sauce is thick and creamy. Add 8 oz (225 g) of the grated cheese and season.

Cook the macaroni to within 1 minute of the cooking time given in the instructions on the packet, drain, rinse in cold water and drain again very thoroughly. Put a layer of the macaroni into a buttered ovenproof dish (or two dishes if you are using the smaller ones), cover with a thick layer of the sauce and mix well. Repeat until the macaroni and sauce have been used up, ending with a layer of sauce.

Mix the remaining grated cheese with the breadcrumbs, crushed cornflakes or Weetabix and sprinkle over the top.

To serve immediately: dot with butter and cook in a moderately hot oven (190°C, 375°F, gas 5) for about 30 minutes until brown and bubbling.

To freeze: allow to cool, then wrap in foil and freeze.

To serve after freezing: thaw overnight in the refrigerator, or for at least 3 to 4 hours at room temperature. Dot with butter and cook as above.

If cooking straight from the freezer, allow 2 to 2½ hours at 190°C, 375°F, gas 5 for the larger casserole, and about 30 minutes less for the smaller ones.

Chocolate Mousse

Serves 10–12

One of the simplest of the old-time favourites.

10 oz (275 g) plain or bitter chocolate
4 tbls (60 ml) brandy (optional)
10 eggs
pinch of salt

¼ pt (150 ml) double or whipping cream (optional)

¼ pt (150 ml) whipping cream (optional)

Break the chocolate into a large bowl, add the brandy, if you are using it, and set over a saucepan of simmering water. Leave to melt, then stir until smooth.

Separate the eggs and stir the yolks, one by one, into the melted chocolate. The mixture will stiffen at first and then become increasingly soft. Stir until absolutely smooth, then remove the bowl from the pan and allow to cool.

Whisk the whites with the salt until they stand in peaks but are not quite dry. Fold gently into the cooled chocolate mixture.

Lightly whip the cream, if you are using it, and fold in also, to make an even richer mousse. Pour into a glass or china serving dish.

To serve immediately: chill in the refrigerator for several hours before serving. Decorate with whipped cream if you wish.

To freeze: cover the bowl with clingfilm and freeze for no longer than a week or two.

To serve after freezing: allow to thaw at room temperature for 2 to 3 hours so that it is still quite chilled.

Variations: instead of the brandy you can use orange curaçao or Grand Marnier, and add a little finely grated orange rind. You can also make a chocolate peppermint mousse by using crème de menthe or a chocolate peppermint liqueur.

Lemon Icebox Pudding

Serves 10–12

Another splendid dessert which looks very decorative for a buffet supper.

8 oz (225 g) butter, preferably unsalted
8 oz (225 g) caster sugar
4 large eggs
finely grated rind and juice of 2 lemons
finely grated rind and juice of 1 orange

1½–2 packets boudoir sponge fingers

¼ pt (150 ml) double or whipping cream (optional)

crystallized orange or lemon slices (optional)

Cream the butter with all but 1 tbls (15 ml) of the sugar until very light and fluffy.

Separate the eggs and beat in the yolks one by one. Add the lemon and orange juice and fold in the finely grated rinds.

Whisk the egg whites until they stand in soft peaks. Add the remaining sugar and whisk until stiff. Fold gently but thoroughly into the yolk mixture.

Line a 2-lb (1-kg) loaf tin with foil (this is easier if you first turn the tin upside down and shape the piece of foil over the tin).

Put a layer of sponge fingers on the bottom of the tin, cover with a layer of lemon mixture and repeat until everything has been used up. Cover with a layer of foil.

To serve immediately: chill in the refrigerator for several hours, or preferably overnight. When you are ready to serve, turn the pudding out of the tin and peel off the foil. Decorate with lightly whipped cream and orange and lemon slices.

To freeze: wrap in foil, then freeze.

To serve after freezing: turn out of the tin, peel off the foil and leave in the refrigerator for 4 to 5 hours. Decorate and serve as above.

Chocolate Refrigerator Cake

Serves 10–12

Quick and easy to make, this is an excellent buffet party dessert, served with whipped cream or ice-cream. It is very rich and should be served quite thinly sliced. You can decorate it with whirls of whipped cream and a few blanched almonds and glacé cherries before serving.

8 oz (225 g) plain or bitter chocolate
2 tbls (30 ml) rum or
   brandy (optional)
2 eggs
1 oz (25 g) caster sugar
8 oz (225 g) unsalted butter

8 oz (225 g) digestive or rich
   tea biscuits, or ginger nuts

whipped cream
blanched almonds
glacé cherries

Melt the chocolate with the rum or brandy, if you are using it, in a bowl set over a saucepan of gently simmering water. Remove the bowl from the pan, stir until smooth and leave to cool.

Beat the eggs with the sugar until very light and fluffy. Blend in the melted chocolate. Warm the butter very gently, until creamy but not runny, and blend into the chocolate mixture. Break the biscuits up coarsely and fold into the mixture.

Line a long, narrow loaf tin with foil and pour in the mixture. Smooth the top.

To serve immediately: leave to set in the refrigerator overnight. Turn out of the tin and peel off the foil before serving. Decorate with whipped cream, blanched almonds and glacé cherries.

To freeze: wrap the tin in foil, then freeze.

To serve after freezing: leave to thaw at room temperature for at least 4 hours, or preferably overnight in the refrigerator. Serve as above.

Note: you can also add some coarsely chopped nuts and/or glacé cherries to the chocolate mixture together with the biscuits.

Byculla Soufflé

Serves 8–10

A rather special dessert for a dinner or buffet party.

1–2 oz (25–50 g) digestive biscuits

6 eggs

7 oz (200 g) caster sugar

½ pt (300 ml) double cream

½ oz (15 g) powdered gelatine

3–4 tbls (45–60 ml) lemon juice

3–5 tbls (45–75 ml) rum

1–2 oz (25–50 g) digestive biscuits

Crush the digestive biscuits and use them to cover the bottom of a 3-pt (1.7-L) soufflé dish.

Separate the eggs and beat the yolks and sugar until thick and fluffy. Whip the cream lightly and fold in.

Sprinkle the gelatine over the lemon juice in a small saucepan and leave for a few minutes until spongy. Heat very gently until the gelatine has melted. Allow to cool slightly, then stir into the yolk mixture, together with the rum, and fold in the egg whites beaten until they stand in soft peaks.

Pour into the soufflé dish.

To serve immediately: chill in the refrigerator for 3 to 4 hours. Sprinkle on the crushed digestive biscuits just before serving.

To freeze: wrap in foil, then freeze.

To serve after freezing: remove from the freezer 5 to 6 hours before you want to eat the soufflé, and leave in the refrigerator. Serve as above.

Iced Coffee Praline Mousse

Serves 10–12

A delicious party dessert, served alone or accompanied by a sauce of puréed raspberries or a raspberry sorbet. The praline is so useful for desserts, and keeps so well in a tin or screw-top jar, that it is worth making a larger amount than the quantities given below and storing for future use, such as Praline Butter (see p. 148).

6 eggs	*Praline*
6 oz (175 g) caster sugar	1½ oz (40 g) sugar
3 tbls (45 ml) very strong coffee (made with 4 tsp (20 ml) instant coffee and the minimum of hot water)	1 tbls (15 ml) water
	3 oz (75 g) blanched almonds
½ pt (300 ml) double or whipping cream	

To make the praline, boil the sugar with the water in a very small saucepan until it turns golden-brown. Remove from the heat and add the blanched almonds, stirring to coat evenly with the syrup. Pour on to a wooden board and leave to cool. Crush the mixture to a powder with a rolling pin or in a blender.

Separate the eggs and whisk the yolks with all but 2 tbls (30 ml) of the sugar until very pale and increased in volume. Whisk in the hot coffee, then fold in the praline, reserving 1 tbls (15 ml) for decoration.

Whip the cream until it stands in soft peaks, then fold into the mixture.

Whisk the egg whites until they stand in soft peaks, then beat in the remaining sugar and continue to whisk until you have a dense meringue. Fold carefully into the coffee mixture, then pour into a glass serving bowl or 2-pt (1.2 L) soufflé dish. Sprinkle on the remaining praline and freeze.

Remove from the freezer 10 to 15 minutes before serving.

Variation: you can make a Chocolate Praline Mousse by grating 4 oz (100 g) plain or bitter chocolate finely and stirring it into the mousse at the same time as the praline.

Cheesecakes

****

Cheesecakes are ideal for large buffet parties, since they can be made well ahead. They freeze well, but should not be left in the freezer for more than two weeks or so.

For some, the perfect cheesecake will be the chewy, cooked cheesecakes of central Europe; for others, the unbaked airy and mousse-like confections to which we have recently become more accustomed, or perhaps the lighter baked cakes, often with fruit toppings, from across the Atlantic.

Lemon Cheesecake (I)

Serves 10–12

This delicate uncooked cake is like a very light lemon mousse set on a crumb base.

Biscuit crumb base
2 oz (50 g) butter

8 oz (225 g) digestive biscuits

Filling
8 oz (225 g) curd cheese
8 oz (225 g) cream cheese
4 oz (100 g) caster sugar
3 large eggs
finely grated rind and juice of
 1 large lemon

½ oz (15 g) powdered gelatine
½ pt (300 ml) double or
 whipping cream

whipped cream
2 oz (50 g) chocolate

Heat the oven to 190°C, 375°F, gas 5.

To make the biscuit crumb base, melt the butter in a small saucepan and add the finely crushed biscuits. Stir well, then press the mixture evenly into a 12-in (30-cm) spring-form tin or a china flan case. Bake in the oven for 10 minutes. Leave to cool.

To make the filling, beat the cheeses with all but 3 tbls (45 ml) of the sugar until smooth. Separate the eggs and beat in the yolks one by one. Add the lemon rind.

Sprinkle the gelatine over the lemon juice in a small saucepan and leave for a few minutes until spongy. Heat very gently until the gelatine has melted. Allow to cool slightly, then beat into the cheese mixture.

Whip the cream until it stands in soft peaks and fold in gently.

Whisk the egg whites until they stand in soft peaks, then slowly add the remaining sugar and whisk until the meringue is stiff and glossy. Fold into the cream cheese mixture.

Pour into the prepared tin or flan case.

To serve immediately: chill in the refrigerator for 3 to 4 hours. Decorate with whipped cream and grated or flaked chocolate.

To freeze: as the filling is very delicate, leave the cake in its tin or flan case, and open-freeze. Wrap with clingfilm and then overwrap with foil.

To serve after freezing: allow to thaw for about 4 hours at room temperature or overnight in the refrigerator. Decorate as above.

Variations:

Chocolate Cheesecake: use 2 tbls (30 ml) rum instead of the lemon rind and juice. Melt 4 oz (100 g) plain or bitter chocolate, beat half into the cheese mixture and dribble the remainder over the cake in a spiral pattern before it sets.

Rum and Raisin Cheesecake: soak 4 oz (100 g) raisins in 2 tbls (30 ml) rum for several hours and fold into the mixture. This recipe is also very nice when made with soft light brown sugar.

Lemon Cheesecake (II)

Serves 10–12

For this cheesecake the filling is exactly the same as for Lemon Cheesecake (I), but instead of digestive biscuits it is set between a light spongy cake.

Cake mixture

4 oz (100 g) butter	5 oz (150 g) self-raising flour
2½ oz (65 g) caster sugar	1 egg

Filling

8 oz (225 g) curd cheese
8 oz (225 g) cream cheese
4 oz (100 g) caster sugar
3 large eggs
finely grated rind and juice of
  1 large lemon

½ oz (15 g) powdered gelatine
½ pt (300 ml) double or
  whipping cream

Heat the oven to 180°C, 350°F, gas 4.

To make the cake mixture, cream the butter with the sugar, then beat in the sifted flour alternately with the beaten egg, a little at a time, making sure that each addition is well mixed in before adding the next. Butter a 9-in (22-cm) spring-form tin and pour in the mixture. Bake in the oven for 30 to 35 minutes, until the sponge is risen, golden and slightly firm to the touch. Turn out on to a wire rack and leave to cool.

To make the filling, beat the cheeses with all but 3 tbls (45 ml) of the sugar until smooth. Separate the eggs and beat in the yolks one by one. Add the lemon rind.

Sprinkle the gelatine over the lemon juice in a small saucepan and leave for a few minutes until spongy. Heat very gently until the gelatine has melted. Allow to cool slightly, then beat into the cheese mixture.

Whip the cream until it stands in soft peaks and fold in gently.

Whisk the egg whites until they stand in soft peaks, then slowly add the remaining sugar and whisk until the meringue is stiff and glossy. Fold into the cream cheese mixture.

Carefully cut the cake in half horizontally, making sure that the top and bottom halves are of equal thickness. Replace the bottom half in the cake tin and pile the filling on top. Top with the other half.

To serve immediately: chill in the refrigerator for 2 to 3 hours before serving.

To freeze: as the filling is very delicate, leave the cake in its tin and open-freeze. Wrap with clingfilm, then overwrap with foil.

To serve after freezing: allow to thaw for about 4 hours at room temperature or overnight in the refrigerator.

Cooked Cheesecake

Serves 8–10

This recipe comes from Germany. It is very quick to make as it needs no crust, and it keeps fresh and moist in the freezer for several weeks.

1 lb (500 g) cream cheese
1 lb (500 g) curd cheese
6 oz (175 g) butter, melted and
  cooled
8 oz (225 g) caster sugar
2 tsp (10 ml) sifted baking powder

4 eggs
4 tbls (60 ml) semolina
finely grated rind and juice of 1 lemon
8 oz (225 g) sultanas (optional)

icing sugar

Heat the oven to 190°C, 375°F, gas 5.

Put all the ingredients into a large mixing bowl and beat until smooth. Pour into an 8–10-in (20–25-cm) loose-bottomed or spring-form cake tin. Bake in the oven for 50–60 minutes, until the cheesecake has risen and the top has lightly browned.

Remove from the oven and leave to cool completely in the tin (the cake will sink and shrink away from the sides of the tin as it cools).

To serve immediately: remove from the tin when cool, place on a cake dish or plate and sprinkle with a little sifted icing sugar before serving.

To freeze: when completely cool, wrap the cake (in its tin if you can spare it) in clingfilm or foil, then freeze. If you cannot spare the tin, open-freeze before over-wrapping.

To serve after freezing: leave to thaw overnight in the refrigerator or for at least 4 to 6 hours at room temperature, before serving as above.

Orange Glazed Cheesecake

Serves 12

This cooked cheesecake is topped with soft fruit to make it look particularly decorative. Ideally it should be glazed on the day you want to eat it, but it can also be frozen with the glaze.

6 oz (175 g) shortcrust pastry
2 lb (1 kg) cream cheese
10 oz (275 g) sugar
2 egg yolks
5 whole eggs

2½ oz (60 g) plain flour
finely grated rind and juice of
  1 orange
¼ pt (150 ml) sour cream

Glaze

3 large oranges	1 tbls sugar
½ pt (300 ml) orange juice	2 tbls (30 ml) orange liqueur
2 tbls (30 ml) cornflour	

Heat the oven to 190°C, 375°F, gas 5.

Roll out the pastry and use to line the base and sides of a 12-in (30-cm) buttered spring-form cake tin. Fill with crumpled foil so that the sides do not collapse and bake blind for 12 to 15 minutes.

Beat the cream cheese with the sugar until smooth. Add the egg yolks and eggs, one by one, beating well between each addition until the mixture is quite smooth. Sift the flour and fold in.

Add the orange rind and juice, and stir in the sour cream.

Pour into the prepared pastry case and return to the oven for 20 minutes, then lower the temperature to 140°C, 275°F, gas 1, and bake for a further 1½ to 1¾ hours, or until a skewer inserted into the centre of the cake comes out clean.

Leave to cool in the tin.

To serve immediately: remove the sides of the tin and chill in the refrigerator for several hours or overnight. Top with the glaze.

Glaze

Peel the oranges carefully, remove all the pith and divide into segments. Stir a little of the orange juice into the cornflour to make a smooth paste, then add the remaining juice and sugar and bring slowly to the boil, stirring well. Simmer for 1 minute, remove from the heat and add the orange liqueur. Allow to cool until lukewarm and beginning to thicken.

Spread a little of the glaze over the cake, arrange the orange segments on top, and carefully spoon over the remaining glaze. Chill in the refrigerator until set.

To freeze: as this cake is fragile, freeze in the tin if possible, well wrapped.

To serve after freezing: allow to thaw at room temperature for 5 to 6 hours, or overnight in the refrigerator. Glaze as above.

Note: the glaze can also be made with canned mandarin segments, but it is best not to freeze these.

Eggs

****

Eggs are one of the few staple foods whose price still fluctuates with the seasons, so it is worth being extravagant with them in spring and early summer, when the price has dropped sharply or your own hens are working overtime.

Whole eggs cannot be frozen in their shells, as these would burst when the contents expanded on freezing, nor can eggs that have been cooked – boiled, poached or fried – on their own. But raw eggs can be frozen out of their shells, either whole or separated, and any number of egg dishes can be stored in the freezer.

To freeze whole eggs, break them up gently with a fork, as you would for making omelettes, then add either a pinch of salt or 1/4 tsp (1.2 ml) sugar per egg. Freeze in small containers and label clearly, stating the date and the number of eggs, and whether salt or sugar has been added. To thaw 4 to 6 eggs, allow about 1 hour at room temperature. Once thawed, use quickly.

These eggs can be used for omelettes, batters, quiches, cakes and custards, or any other dishes to which you would normally add whole eggs.

However, it is generally better to freeze the yolks and whites separately, especially as it is not always convenient when cooking to use both parts, so that, for instance, you frequently have whites left over when making mayonnaise, or yolks when making meringues. It is then particularly useful to be able to freeze the left-over half, as neither yolks nor whites keep well in the refrigerator for more than a few days.

Do not keep eggs in the freezer for more than 3 months.

It is best not to freeze dishes made with frozen eggs.

EGG YOLKS

To freeze yolks, break them up with a fork and add a pinch of salt or ¼ tsp (1.2 ml) sugar per yolk.

Freeze several yolks together in small containers, labelling with the date and number, and stating whether salt or sugar has been added.

Yolks can also be frozen in ice cube trays, then stored in polythene bags, so that you can take out one yolk at a time. If you have stirred several yolks together but want to freeze them separately in this way, remember that the yolk from a medium to large egg equals roughly 1 scant tbls (15 ml).

Do not store in the freezer for more than 3 months.

Frozen yolks can if necessary be used straight from the freezer if you want to thicken soups and sauces. Add when the saucepan has been taken off the heat, and stir constantly until the yolk has completely thawed and been amalgamated into the sauce.

For other dishes, thaw thoroughly and allow to reach room temperature first (about 1 hour for 4 to 6 yolks). Stir frequently while thawing, to prevent lumps or skin from forming, and use quickly once thawed.

Cheese Straws

Best eaten on the day they are made, when they are at their light and flaky best, cheese straws are quick to make, and are nowadays an unusual accompaniment to pre-meal drinks. A good, well-flavoured cheddar cheese should be used.

4 egg yolks	4 oz (100 g) butter or margarine
6 oz (175 g) grated cheese	freshly ground pepper
6 oz (175 g) plain flour	

Frozen yolks must be allowed to thaw thoroughly. Stir several times while they are thawing.

Heat the oven to 180°C, 350°F, gas 4.

Mix together the cheese and flour. Rub in the butter or margarine and stir in the beaten egg yolks. Season generously with pepper (you will probably not need any salt as the cheese will be salty). Mix to a stiff paste.

Roll out on a lightly floured board or work surface ⅛ inch (2.5 mm) thick and cut into fingers about 3 × ½ in (7 × 1 cm). Arrange on a floured baking sheet and bake for 15 to 20 minutes until light golden colour; watch carefully to see that they do not get too brown.

If the straws are not to be eaten straight away, store in an airtight tin and warm gently before serving.

Gnocchi with Cheese Sauce

Serves 6 as a starter, 4 as a main course

2 egg yolks
2 pts (1.2 L) milk
4 oz (100 g) semolina, preferably
 a coarse variety
salt and freshly ground pepper
freshly grated nutmeg
2 oz (50 g) butter

2 oz (50 g) plain flour
6 oz (175 g) grated mature cheddar
 cheese

2 oz (50 g) grated parmesan cheese
1 oz (25 g) butter

Frozen yolks must be allowed to thaw thoroughly. Stir several times while they are thawing.

Bring half the milk to the boil in a saucepan and gradually sprinkle in the semolina, stirring. Continue to cook, stirring, over a gentle heat for a few minutes, until the mixture thickens and comes away from the side of the pan. Season with salt and pepper, and generously with nutmeg, which loses some of its flavour in cooking. Remove from the heat and allow to cool slightly, then beat in the egg yolks. Turn on to an oiled board and leave to cool.

Heat the oven to 180°C, 350°F, gas 4.

Melt the butter in a saucepan, add the flour and stir for 2 to 3 minutes over a gentle heat without allowing it to colour. Stir in the remaining milk and when the sauce is thick and smooth add most of the grated cheese. Season with salt, pepper and nutmeg. Leave over the lowest possible heat while you make the gnocchi.

Butter one or two shallow ovenproof dishes and with well-floured hands shape the semolina mixture into balls about 1 in (2.5 cm) in diameter. Arrange them in a single layer in the dishes and sprinkle with the remaining cheese. Cover with the sauce.

To serve immediately: sprinkle with the parmesan cheese and dot with the butter. Bake in the oven for about 30 minutes until bubbling and golden-brown. If necessary, finish by putting under a hot grill for a minute or two to brown the top.

To freeze: allow to cool, then wrap the dishes well in foil and freeze.

To serve after freezing: either allow to thaw and reheat as above; or transfer straight from the freezer into a hot oven (200°C, 400°F, gas 6), having first sprinkled on the parmesan cheese and dotted with butter. After 30 minutes turn the heat down to 180°C, 350°F, gas 4, and cook for a further 45 minutes or so.

Zabaglione

Serves 4

This dish is said to have been created in desperation one day by the chef to the Piedmontese Duke of Savoy, Emmanuel Philibert, and in thanksgiving he called it after the patron saint of cooks, Saint Jean Bayon.

4 egg yolks   4 tbls (60 ml) marsala or sweet sherry
4 oz (100 g) caster sugar

Frozen egg yolks must be allowed to thaw thoroughly. Stir several times while they are thawing.

Beat the yolks with the sugar in a bowl until very thick and pale. Slowly beat in the marsala or sherry. Place the bowl over a saucepan of simmering water or transfer the mixture to the top of a double boiler. If you have a heavy copper or brass pan, however, you can cook the zabaglione over direct heat.

Beat with a balloon whisk over a moderate heat, and as the mixture rises, but just before it comes to the boil, pour into 4 tall wine glasses, preferably long-stemmed, with sugar-frosted rims.

Serve with *langues de chat* biscuits or sponge fingers.

Variation: you can make zabaglione into a rich ice cream by allowing it to cool and then folding in ¼ pt (150 ml) lightly whipped double or whipping cream before freezing. But in this case do not use frozen egg yolks.

Another deliciously extravagant variation for special occasions is to substitute champagne for marsala, using a bare ¼ pt (120 ml) champagne to 4 egg yolks. This is the perfect accompaniment to any summer fruit dessert.

Crème Nostradamus

Serves 4–6

Another dessert with an exotic history, since it is said to have been invented by the celebrated astrologer and physician to strengthen convalescents after a plague epidemic.

4 egg yolks	½ pt (300 ml) milk
2 tbls (30 ml) clear honey	1–2 tbls (15–30 ml) brandy
1 oz (25 g) blanched almonds	1 oz (25 g) butter
1 tbls (15 ml) cornflour or potato flour	

Frozen egg yolks must be allowed to thaw thoroughly. Stir several times while they are thawing.

Beat the yolks with the honey in a bowl until thick and foamy. Grind the almonds finely and mix them with the cornflour or potato flour. Beat into the honey mixture.

Bring the milk to the boil in a saucepan and pour slowly on to the mixture. Return to the saucepan and stir over a gentle heat until the mixture thickens, but do not allow it to boil. Remove from the heat. Slowly stir in the brandy and then the butter, cut into small pieces.

Pour into individual serving glasses and serve warm with *langues de chat* biscuits or sponge fingers.

Crème Brûlée

Serves 4–6

This rich, smooth dessert is said to have originated at Trinity College, Cambridge. Though more often found on restaurant menus, it is quite simple to make at home.

6 egg yolks	2 tsp (10 ml) caster sugar
1 pt (600 ml) double or whipping cream	4–5 oz (100–150 g) soft light brown sugar
vanilla pod or 2–3 drops vanilla essence	

Frozen egg yolks must be allowed to thaw thoroughly. Stir several times while they are thawing.

Add about ¼ pt (150 ml) of the cream to the yolks in a bowl and stir until smooth. Heat the remaining cream with the vanilla pod in a heavy non-stick saucepan, and when it reaches simmering point remove from the heat and stir slowly into the yolks and cream mixture.

Add the caster sugar, return to a gentle heat and stir constantly until the custard thickens sufficiently to coat the back of the spoon. On no account allow to reach simmering point.

Remove from the heat and remove the vanilla pod or add vanilla essence. Strain into individual ramekin dishes and leave to set in the refrigerator for several hours or overnight.

A few hours before serving chill for 30 minutes or so in the freezer, then sprinkle with the soft brown sugar, making sure that the custards are completely covered with an even layer of sugar.

If you have a salamander, use it to caramelize the tops. If not, heat the grill to maximum heat, then quickly place the creams underneath, as close as possible to the element without touching, and leave until the sugar melts and bubbles. Chill in the refrigerator again for at least 1 hour before serving; the tops will have set into a hard caramel lid, to be broken with a spoon before eating.

Lemon Biscuits

The uncooked dough can be kept in the freezer, ready to bake at very short notice, or in the refrigerator. But if you have used frozen yolks do not keep in the refrigerator for more than a day or two.

4 egg yolks
2 oz (50 g) butter
6 oz (175 g) sugar
finely grated rind and juice of
  1 lemon

1 tsp (5 ml) baking powder
6 oz (175 g) plain flour

Frozen egg yolks must be allowed to thaw thoroughly. Stir several times while they are thawing.

Cream the butter with the sugar just enough to make a smooth mixture. Add the lemon rind and juice and beat in the egg yolks. Blend the baking powder into the flour and mix in. Or put all the ingredients together into a food processor and process until the dough gathers itself into a ball.

Put the dough on to a sheet of clingfilm and shape into a roll about 1½ in (3.5 cm) in diameter. Wrap, then refrigerate or freeze.

When you are ready to bake the biscuits, heat the oven to 190°C, 375°F, gas 5. Slice the dough into rounds about ¼ in (5 mm) thick (this can be done as soon as the mixture comes out of the freezer) and space well apart on greased baking sheets.

Bake for 8 to 10 minutes; 12 to 15 minutes if you are cooking from frozen.

EGG WHITES

Egg whites need no preparation before storing in the freezer. Simply put them into small containers and label with the date and number of whites.

They may also be frozen individually in ice cube trays (1 white will usually make 2 cubes) and stored in polythene bags, so that they can be used singly – for soufflés, for instance, which generally need more whites than yolks.

Egg whites can be stored in the freezer for 6 months.

If the whites are to be whipped after freezing it is important to keep them absolutely dry, so, if some ice crystals have formed on the top, scrape these off thoroughly before leaving the whites to thaw.

Frozen whites, once thawed, can be whipped very successfully, and they are excellent for meringues, as they whip into a particularly dense meringue mixture.

When the whites have thawed, use them quickly, and do not freeze dishes containing frozen egg whites unless they have been cooked.

Asparagus Mousse

Serves 4–5

A light and delicate cold starter. Home-made mayonnaise should be used if possible, but a good bought variety will do quite well instead.

2 egg whites	4 tbls (60 ml) mayonnaise
2 tsp (10 ml) powdered gelatine	8¾ oz (250 g) can asparagus
scant ¼ pt (125 ml) chicken stock	

Frozen egg whites must be allowed to thaw to room temperature.

Sprinkle the gelatine over the chicken stock and heat gently, stirring, until it has dissolved. Allow to cool, then mix well with the mayonnaise.

Drain the asparagus and reserve a few of the tips for the garnish. Chop the remainder, not too finely, and add to the mayonnaise mixture.

Whisk the egg whites until they stand in soft peaks and fold in.

Divide the mixture between 4 or 5 ramekins and chill in the refrigerator for 2 to 3 hours, until set. Just before serving, decorate with the reserved asparagus tips.

Cheese Pudding

Serves 6–8

An excellent dish for unexpected visitors or hungry children, especially if you have grated cheese and breadcrumbs ready to hand in the freezer.

1 egg white	a little made mustard
4 whole eggs	freshly grated nutmeg
1 pt (600 ml) milk	salt and freshly ground pepper
8 oz (225 g) grated cheddar cheese	
5 oz (150 g) coarse fresh breadcrumbs, brown or white	

Frozen egg whites must be allowed to thaw to room temperature.

Heat the oven to 220°C, 425°F, gas 7.

Beat together the whole eggs and the egg white and add the milk. Stir in the cheese and breadcrumbs, a little made mustard and a few gratings of nutmeg. Season.

Pour into a well-greased ovenproof dish and cook for 25 to 30 minutes, until golden and firm to the touch. Serve at once.

SOUFFLÉS

A soufflé is extremely easy to make, yet whenever it is brought to table, with the lightly browned, crusted top delicately quivering above the dish, it raises a cry of admiration. The only problematical thing about making a soufflé for guests is the last-minute preparation and the timing, but with the aid of a freezer this need cause no difficulty; the soufflé can be fully prepared ahead of time, frozen uncooked, and placed straight from the freezer into the oven. Prepared in this way, a soufflé is best made only a day or two ahead, but can be stored in the freezer for up to a week.

The soufflé dish must be generously buttered before the mixture is poured in. For savoury soufflés it can also be dusted with fine dry breadcrumbs or finely grated dry cheese, or for sweet soufflés with caster sugar or ground almonds.

To make the soufflé look truly impressive, use a dish that will only just hold the mixture before it is cooked, so that in baking it will rise well above the rim of the dish. To do this make a collar from greaseproof paper or foil 3–4 in (7–10 cm) wide and long enough to allow a generous overlap. Butter it well and tie tightly with string just below the rim of the soufflé dish, where there is usually a slight overhang for this purpose. Remove the collar just before bringing the soufflé to the table.

Cheese Soufflé

The simplest and most basic of all the soufflés, but no less delicious for that, this makes an economical but delicate supper dish, or an excellent starter for a dinner party. The quantities given below are sufficient for 1 × 2-pt (1.2-L) soufflé dish, which will serve 4 to 6, or else for 2 × 1-pt (600-ml) dishes, each of which will serve 2 to 3, so that you can make one and freeze one.

1 egg white

4 oz (100 g) butter

2 oz (50 g) plain flour

1 pt (600 ml) milk

4 whole eggs

4 oz (100 g) finely grated cheese,
    preferably cheddar or gruyère

salt and freshly ground pepper

The frozen egg white must be allowed to thaw to room temperature.

For serving immediately, heat the oven to 190°C, 375°F, gas 5.

Melt the butter in a saucepan, add the flour and cook over a moderate heat for 3 to 4 minutes, stirring well, without allowing the mixture to brown. Gradually add the milk and stir well, until thick and smooth. Remove from the heat.

Separate the whole eggs and add the yolks one by one to the sauce, stirring until each is amalgamated before adding the next. Stir in all but 1 tbls (15 ml) of the grated cheese and season to taste. Use the remaining cheese to prepare the soufflé dish (see p. 124).

Whisk the whites until they stand in soft peaks, then fold them gently into the cheese mixture. Pour into the prepared soufflé dish.

To serve immediately: bake for 40 to 50 minutes, until the top has risen and is golden-brown but the centre is still creamy. Bring straight to the table.

To freeze: wrap the dish in foil, then freeze.

To serve after freezing: heat the oven to 220°C, 425°F, gas 7, and transfer the soufflé straight from the freezer to the centre of the oven. Bake for 20 minutes, then lower the heat to 190°C, 375°F, gas 5, and bake for a further 30 to 40 minutes. Serve at once.

Note: the soufflé can also be baked in individual ramekin dishes (this quantity makes about 16). To serve immediately, they will need only 10 to 15 minutes in the very hot oven, and an extra 10 to 15 minutes after the oven has been turned down if they have been frozen.

Vegetable Soufflé

More unusual than a cheese soufflé, and especially pretty if you use a green or orange vegetable such as spinach or carrot. But this soufflé is also excellent – and more intriguing – made with a purée of a white vegetable such as Jerusalem artichoke or cauliflower. It is especially good served with a creamy, lightly cheese-flavoured hollandaise or béchamel sauce. Whatever vegetable you use, it should be cooked very lightly and well drained before being puréed. Flavour the soufflé mixture according to the vegetable you are using: for instance, a touch of nutmeg enhances the flavour of spinach, coriander goes well with carrots, a little sugar with turnips, and a touch of ginger is excellent with cauliflower.

The quantities given below are sufficient for 1 × 2-pt (1.2-L) soufflé dish, which will serve 4 to 6, or else for 2 × 1-pt (600-ml) dishes, each of which will serve 2 to 3, so that you can cook one and freeze one.

2 egg whites	2 oz (50 g) grated cheese (optional)
¼ pt (150 ml) vegetable purée	salt and freshly ground pepper
4 oz (100 g) butter	herb or spice (see above)
2 oz (50 g) plain flour	4 whole eggs
¾ pt (450 ml) milk	

The frozen egg whites must be allowed to thaw to room temperature.

For serving immediately, heat the oven to 190°C, 375°F, gas 5.

Gently heat the vegetable purée.

Melt the butter in a saucepan, add the flour and cook over a moderate heat for 2 to 3 minutes, stirring well. Add the vegetable purée and stir until smooth. Gradually add the milk, stir until smooth, then remove from the heat. Stir in the cheese, seasoning and flavouring.

Separate the whole eggs and break up the yolks with a fork. Stir the yolks into the sauce a little at a time. Whisk the 6 whites until they stand in soft peaks, then fold gently into the mixture. Pour into the prepared soufflé dish.

To serve immediately: bake in the oven for 30 to 50 minutes, until the top has risen and is golden-brown, but the centre is still creamy. Bring straight to the table.

To freeze: wrap the dish in foil, then freeze.

To serve after freezing: heat the oven to 220°C, 425°F, gas 7, and transfer the soufflé straight from the freezer to the centre of the oven. Bake for 20 minutes, then lower the heat to 190°C, 375°F, gas 5, and bake for a further 30 to 40 minutes until the soufflé has risen and has a golden brown crust on top, but is still creamy at the centre. Serve at once.

Note: the soufflé can also be baked in individual ramekin dishes (this quantity makes about 16). To serve immediately, they will need only 10 to 15 minutes in the very hot oven and an extra 10 to 15 minutes after the oven has been turned down if they have been frozen.

Sour Cream Soufflé

Serves 6

A light and delicate starter which is simplicity itself to make.

1 egg white	¾ oz (20 g) grated gruyère or
¼ pt (150 ml) sour cream	parmesan cheese
1½ oz (40 g) plain flour	2 whole eggs
1 tbls (15 ml) finely chopped chives	salt and freshly ground pepper

The frozen egg white must be allowed to thaw to room temperature.

Heat the oven to 180°, 350°F, gas 4.

Mix together the sour cream and the flour, until smooth. Add the chives and cheese. Separate the eggs and add the well-beaten yolks. Season with salt and pepper. Stir well to mix.

Whisk the 3 egg whites until they stand in soft peaks, then fold into the mixture. Divide among 6 ramekin dishes (the mixture should come to just below the rim).

To serve immediately: put the ramekin dishes into a roasting tin and pour in hot water to come about ½ in (1 cm) up the sides of the dishes. Cook for about 15 minutes, until the soufflés have risen well and are slightly firm to the touch. Serve at once.

To freeze: wrap well in foil, then freeze.

To serve after freezing: transfer the ramekin dishes straight from the freezer to a roasting tin containing about ½ in (1 cm) hot water. Bake in a hot oven (220°C, 425°F, gas 7) for 10 minutes. Turn the oven down to 180°C, 350°F, gas 4 and bake for a further 5 minutes, until the soufflés have risen well and are slightly firm to the touch. Serve at once.

St Clement's Soufflé

Serves 6–8

This sweet soufflé, placed in the oven at the beginning of the meal, can be produced like a rabbit out of a hat at dessert-time. If you are cooking it straight from the freezer, make it in ramekin dishes, as one big soufflé would take too long to cook. For extra effect, sprinkle some finely chopped candied orange and lemon peel at the bottom of the soufflé or ramekin dishes before spooning in the soufflé mixture.

2 egg whites
4 whole eggs
4 oz (100 g) caster sugar
finely grated rind and juice of
  ½ orange
finely grated rind and juice of
  ½ lemon

2 tbls (30 ml) orange curaçao or
  Grand Marnier

icing sugar
single cream (optional)

Frozen egg whites must be allowed to thaw to room temperature.

For serving immediately, heat the oven to 190°C, 375°F, gas 5.

Separate the whole eggs and whisk the yolks with half the sugar until very pale and fluffy. Slowly whisk in the orange and lemon juice and the liqueur. Add the finely grated orange and lemon rind.

Whisk the 6 egg whites until they begin to foam, then slowly add the remaining sugar and continue to whisk until they stand in soft peaks. Fold into the yolk mixture and pour into the prepared soufflé or ramekin dishes.

To serve immediately: bake in the oven for 40 to 50 minutes for one large soufflé, or 20 to 25 minutes for individual ramekin dishes. The top should have risen and be golden-brown, but the centre should still be creamy. Sprinkle with sifted icing sugar and bring straight to the table. Hand cream separately if liked.

To freeze: wrap the ramekin dishes in foil, then freeze.

To serve after freezing: heat the oven to 220°C, 425°F, gas 7 and transfer the soufflés straight from the freezer to the centre of the oven. Bake for 10 minutes, then lower the heat to 190°C, 375°F, gas 5 and bake for a further 10 to 15 minutes, until the soufflés have risen. Serve as above.

Apple Snowballs

Serves 4

An unfailing favourite with children. For adults you could stuff the cavities of the apples with raisins or glacé fruit macerated in kirsch or any other fruit brandy or liqueur. Cooking or eating apples may be used.

2 egg whites
4 medium apples
4 cloves (optional)
4 tbls (60 ml) caster sugar

2 oz (50 g) flaked almonds (optional)

double or whipping cream
grated chocolate (optional)

Frozen egg whites must be allowed to thaw to room temperature.

Heat the oven to 150°C, 300°F, gas 2.

Halve, peel and core the apples. Place them cut side down in a buttered shallow baking dish, spaced well apart, and stick a clove, if you are using them, into each dome.

Whisk the egg whites until they stand in soft peaks, then very slowly whisk in the caster sugar until stiff and glossy.

Spoon the meringue mixture over each apple to make separate snowballs, sprinkle with the flaked almonds, if you are using them, and bake in the oven for 30 to 40 minutes, until the apples are just tender and the meringue is crisp on the outside.

Serve with lightly whipped cream if you like, and for a special treat sprinkle with a little grated chocolate.

White Chocolate Mousse

Serves 4

This is simplicity itself to make, and children love it. For adults, pour a little liqueur, brandy or rum over each helping before serving. It is best to whisk the whites by hand with a balloon whisk in order to achieve the necessary density. Otherwise use a powerful electric mixer with a balloon whisk. Do not make the mousse more than 1 to 2 hours before serving it.

4 egg whites
3½ oz (90 g) plain or bitter chocolate
4 tbls (60 ml) caster sugar
½ tsp (2.5 ml) ground cinnamon
(optional)

1 tbls brandy or other liqueur (optional)
single cream (optional)

Frozen egg whites must be allowed to thaw to room temperature.

Grate the chocolate very finely.

Whisk the egg whites until they stand in soft peaks, then whisk in the sugar until the mixture is very stiff and glossy. If you are using cinnamon, blend it into the sugar before whisking in. Fold in the grated chocolate. Pour into individual serving glasses and refrigerate.

You can pour a little liqueur round the edges and mask the mousse with cream just before serving.

Note: you can make a completely white mousse by using white chocolate. This has no cocoa solids, so it does not have as distinctive a taste, but it is amusing to see whether your guests can identify the flavour.

Rosy Mousse

Serves 4

You can make this mousse with freshly puréed fruit, with frozen fruit purée, or even with a good home-made strawberry or raspberry jam. Do not make more than 1 to 2 hours before serving.

4 egg whites
2 tbls (30 ml) kirsch or
  orange liqueur
4 tbls (60 ml) fruit purée or
  sieved jam

finely grated rind of ½ lemon
2 oz (50 g) blanched almonds

Frozen egg whites must be allowed to thaw to room temperature.

Stir the kirsch or liqueur into the fruit purée or jam.

Whisk the whites until they stand in very stiff peaks, then slowly but thoroughly fold in the fruit purée or jam together with the lemon rind. Pile into individual serving glasses.

Roast or toast the almonds, chop coarsely or cut into slivers, and scatter them over each glass.

Snow Pudding

Serves 6–8

Improbable as this recipe from America looks to English eyes, it is extremely good, and popular with adults and children alike. You can vary the accompaniment to suit different tastes.

4 egg whites
½ oz (15 g) powdered gelatine or
  1 leaf gelatine
¼ pt (150 ml) water
4 oz (100 g) caster sugar
finely grated rind and juice of
  1 large lemon

4 tbls (60 ml) finely crushed digestive
  biscuits, macaroons or praline (see
  p. 110)

Frozen egg whites must be allowed to thaw to room temperature.

Sprinkle the gelatine over 4 tbls (60 ml) of the water and leave for a few minutes. Bring the remaining water to the boil, remove from the heat, add half the sugar and the softened gelatine and stir until the gelatine and sugar are completely dissolved and the liquid is clear. Add the lemon rind and juice. Leave to cool until just beginning to set, then whisk rapidly with a wire whisk until thick and frothy.

Whisk the egg whites until they stand in soft peaks, beat in the remaining sugar and fold into the gelatine mixture. Whisk together until really stiff, then pour into a rinsed-out shallow dish or tin with straight sides. Chill in the refrigerator for 2 to 3 hours, until set.

Sprinkle with the crushed biscuits or macaroons or the praline, and cut into squares to serve.

Serve with chocolate sauce, fruit purée or ice-cream.

Apricot Shortcake

Serves 4–5

This dessert comes from Romania, using (in its native habitat) jam made from the luscious apricots which grow there so plentifully.

1 egg white	4 oz (100 g) caster sugar
6 oz (175 g) plain flour	1 whole egg
4 oz (100 g) butter or margarine	about 8 oz (225 g) apricot jam

The frozen egg white must be allowed to thaw to room temperature.

Heat the oven to 200°C, 400°F, gas 6.

Sift the flour into a mixing bowl. Add the diced butter or margarine and rub in with the fingertips. Stir in half the sugar. Separate the egg and add the yolk. Knead until the mixture comes away from the side of the bowl. Press into a floured 8-in (20-cm) flan tin and bake in the centre of the oven for 25 to 30 minutes until the shortcake is light brown and firm to the touch. Remove from the oven and leave to cool in the tin for about 20 minutes.

Spread the shortcake generously with apricot jam. Whisk the 2 egg whites until they stand in stiff peaks and fold in the remaining caster sugar. Spread over the jam. Return to the hot oven for 10 minutes.

Serve warm or cold.

MERINGUES

Meringues are said to have originated in Nancy at the court of King Stanislaus of Lorraine. Whoever invented them was a benefactor of mankind, for they are a marvellous basis for all kinds of desserts, not least because they freeze excellently. The only snag is that they can be quite tricky to make, and even when you have been turning them out successfully for years they may suddenly 'go sour' on you and refuse to cook properly. Sometimes, if the top is crisp but the base is still soft, it helps to turn them carefully upside down and continue to cook until the base too is crisp. Meringues can be frozen even if they have been made with frozen egg whites.

4 egg whites	8 oz (225 g) caster sugar
¼ tsp (1.2 ml) cream of tartar	

Frozen egg whites must be allowed to thaw to room temperature.

Heat the oven to 100°C, 200°F, gas ¼.

Whisk the egg whites until they stand in stiff peaks, add the cream of tartar and whisk again. Then very slowly add the sugar, whisking steadily all the time and being careful to add only a very little at once.

Lightly oil sheets of foil placed on baking sheets or tins. Either pipe the mixture on to the foil to make small meringues or use a dessertspoon to make neat ovals, finishing off with a twirl of the spoon. Place in the lower part of the oven and bake for 1½ to 3 hours. After the first hour turn the oven down, if the meringues are starting to colour.

When the meringues are firm to the touch and can be easily lifted off the foil, remove to a wire rack.

To freeze: allow to cool. As meringues are very fragile, they should be stored in the freezer in tins or cardboard or plastic boxes, or else wrapped very carefully and not moved about in the freezer.

To serve after freezing: meringues can be served straight from the freezer.

Variation: you can also shape the meringue mixture into large circles or 'nests', for filling later on, though these are trickier to keep intact in the freezer.

Meringues with Chestnut

Serves 8

4 egg whites
¼ tsp (1.2 ml) cream of tartar
8 oz (225 g) caster sugar
8 oz (225 g) can sweetened chestnut
  purée

½ pt (300 ml) double or
  whipping cream

marrons glacés (optional)

Frozen egg whites must be allowed to thaw to room temperature.

With the egg whites, cream of tartar and sugar, make three meringue circles as described in the previous recipe. Or use meringue circles from the freezer.

Lay one of the circles on a round serving dish and spread with half the chestnut purée. Lay a second circle on top and spread with the remaining purée. Lay the third circle on top of this and cover with the cream whisked until it stands in soft peaks.

Decorate with marrons glacés if you can spare them! If you go to Italy in the autumn, it is always worth looking round for shops which sell broken pieces, as these are considerably cheaper than the perfect ones.

Hazelnut Meringues

4 egg whites
6 oz (175 g) unblanched hazelnuts
¼ tsp (1.2 ml) cream of tartar
8 oz (225 g) caster sugar

finely grated rind of 1 lemon

½ pt (300 ml) double or whipping cream

Frozen egg whites must be allowed to thaw to room temperature.

Heat the oven to 100°C, 200°F, gas ¼.

Roast the hazelnuts lightly in the oven for about 20 minutes. Without removing the skins, put them through a coarse mouli or chop finely.

With the egg whites, cream of tartar and sugar, make the meringue mixture as described on p. 132. Fold in the hazelnuts and the lemon rind. Shape into small meringues, lay on lightly oiled baking sheets or foil and bake for 2 to 3 hours, until the meringues are firm but not coloured. Cool on a wire rack.

To serve immediately: serve sandwiched with lightly whipped cream. They are particularly good eaten with raspberries or with fresh raspberry sauce.

To freeze: store carefully in the freezer (see p. 132).

To serve after freezing: the meringues can be served straight from the freezer. Serve as above.

Variation: walnuts or pecans can be substituted for the hazelnuts. Or you can use ground almonds, though the meringues will have a blander and less interesting texture.

Macaroons

These little macaroons are made in a trice, and do not have the bitterness of artificial almond flavouring which most bought ones suffer from nowadays. They go well with ice cream and many other delicate cold desserts.

1 large egg white	6–7 drops almond essence
4 oz (100 g) ground almonds	rice paper
8 oz (225 g) caster sugar	

The frozen egg white must be allowed to thaw to room temperature.

Heat the oven to 180°C, 350°F, gas 4.

Mix together the ground almonds, sugar and almond essence. Stir in the lightly beaten egg white until the mixture is a smooth paste. With floured hands shape the paste into about 24 small balls roughly the size of marbles, and flatten them slightly.

Lay the rice paper on baking sheets and place the macaroons on top, leaving a little space between them as they will spread during baking. Bake in the oven for 10 to 15 minutes. They should be no darker than the lightest golden, for if they cook too long they will become hard.

Remove from the oven and leave on the baking sheets until cold, then trim away the excess rice paper. Store the macaroons in an airtight tin. They also keep well in the freezer.

White Cookies

Halfway between macaroons and rock cakes, these cookies are very quick and simple to make. If possible, use whole pieces of either lemon or orange candied peel and chop it yourself – it gives a much better flavour than mixed cut peel from packets.

4 egg whites	pinch of salt
4 oz (100 g) candied peel	3 oz (75 g) ground almonds
4 oz (100 g) unsalted butter	6 oz (175 g) plain flour
8 oz (225 g) caster sugar	

Frozen whites must be allowed to thaw to room temperature.

Heat the oven to 180°C, 350°F, gas 4.

Chop the candied peel finely. Cream the butter with half the sugar until very light and fluffy. Whisk the whites with the salt until they stand in soft peaks, then slowly whisk in the remaining caster sugar until thick and glossy. Fold gently into the butter and sugar.

Blend the ground almonds with the flour and fold into the mixture, together with the candied peel.

Place teaspoonfuls of the mixture on greased baking sheets, leaving plenty of space for the cookies to spread during baking.

Bake for 15 to 20 minutes, until light golden-brown.

Silver Cake

This is a very light sponge made with egg whites and only a little butter. It is delicious either on its own or used as the basis for desserts which call for sponge cake. It keeps very well stored in an airtight tin or in the freezer.

4 egg whites	1 tbls (15 ml) baking powder
3 oz (75 g) butter	6 tbls (90 ml) milk
7 oz (200 g) caster sugar	½ tsp (2.5 ml) almond essence
8 oz (225 g) self-raising flour	

Frozen egg whites must be allowed to thaw to room temperature.

Heat the oven to 180°C, 350°F, gas 4.

Butter and lightly flour an 8-in (20-cm) cake tin.

Cream the butter with the sugar until light and fluffy. Sift the flour with the baking powder and add alternately with the milk. Whisk the egg whites until they stand in peaks and fold into the mixture, together with the almond essence.

Pour into the prepared cake tin and bake in the centre of the oven for about 45 minutes, until the sponge is well risen, golden-brown and firm to the touch. Turn out on to a wire rack and leave to cool completely.

To freeze: as soon as the cake is cold, wrap and freeze.

To serve after freezing: thaw at room temperature for about 4 hours.

Angel Cake

Lighter even than the Silver Cake, this is an excellent way of using up vast quantities of egg whites. Serve with soft fruit or fruit salad, or cover with whipped cream or a very light chocolate fudge icing.

10 egg whites	1 tbls (15 ml) lemon juice
1 oz (25 g) cornflour	1 tsp (5 ml) cream of tartar
3 oz (75 g) plain flour	4 oz (100 g) caster sugar
4 oz (100 g) icing sugar	finely grated rind of ½ lemon
½ tsp (2.5 ml) salt	

Frozen egg whites must be allowed to thaw to room temperature.

Heat the oven to 180°C, 350°F, gas 4.

Lightly dust a 10-in (25-cm) spring-form tin with a central funnel with a little flour and icing sugar. Shake off the excess.

Sift the cornflour, flour and icing sugar together several times.

Divide the egg whites between two large, scrupulously clean and dry bowls. Add half the salt and lemon juice to each bowl and whisk each half separately, preferably with a balloon whisk. You can also use a rotary whisk or the balloon whisk of an electric mixer, but nothing is so good as a hand-held balloon whisk, which allows you to incorporate the maximum amount of air. When the whites start to froth, add half the cream of tartar and continue to beat until they begin to stand in peaks. Slowly beat in half the caster sugar and lemon rind, and continue to beat until the mixture stands in firm glossy peaks.

Repeat with the remaining egg whites.

Combine the two batches of meringue mixture in a large bowl, or spread on a large, flat dish or clean working surface. Sift some of the flour mixture over the top and lightly fold in with a metal spoon or a spatula. Continue until all the flour has been used up.

Spoon the mixture into the prepared cake tin, run a knife through the centre of the mixture to expel any large air bubbles and smooth the top.

Bake in the centre of the oven for 45 to 50 minutes, or until the surface is firm enough to spring back when lightly pressed with the fingertip. Remove from the oven, invert over a wire rack and leave to cool thoroughly before removing the tin.

The cake is so soft and springy that it is better to pull it into portions than to cut it with a knife.

Carolina Cookies

These delicately spiced biscuits are incredibly light, and are wonderful
with ices, sorbets or fruit compotes. Keep a supply in the freezer, and
you will always have some on hand to serve with desserts at short notice,
as they take virtually no time to thaw.

2 egg whites	4 tsp (20 ml) ground cinnamon
1 lb (450 g) unsalted butter or margarine	2 tsp (10 ml) ground ginger
	pinch of salt
12 oz (375 g) soft light brown sugar	1 ¼ lb (550 g) plain flour

Frozen egg whites must be allowed to thaw to room temperature.

Heat the oven to 190°C, 375°F, gas 5.

Cream the butter or margarine with the sugar until very light and
fluffy. Sift the spices and salt with the flour. Beat the egg whites into the
creamed mixture until smooth and very light. Fold the flour into the
mixture and beat briefly until smooth.

Butter baking sheets and spoon on the mixture. Pat out quite thinly.
Bake in the oven for 15 to 20 minutes, until light golden-brown.

Leave to cool for a few minutes, then cut into diamond shapes and
cool on wire trays.

To freeze: freeze in plastic boxes or tins.

To serve after freezing: the biscuits may be served straight from the
freezer.

Cinnamon Stars

These traditional German Christmas biscuits not only taste delicious
but are also very pretty, and can even be used as Christmas tree
decorations. Their flavour is best if you use whole, unblanched almonds
and blanch and grind them yourself, but this refinement is not essential.

2 egg whites	½ oz (15 g) ground cinnamon
5 oz (150 g) caster sugar	pinch of ground cloves
5 oz (150 g) ground almonds (see above)	finely grated rind of ½ lemon

Icing

2 egg whites	juice of ½ lemon
6 oz (175 g) icing sugar	some *nonpareils*

Frozen egg whites must be allowed to thaw to room temperature.

Whisk the egg whites with the caster sugar until very thick and bulky. Blend the cinnamon and cloves into the ground almonds, add the lemon rind and fold into the meringue mixture. Gather into a ball and chill in the refrigerator for at least 30 minutes.

Heat the oven to 190°C, 375°F, gas 5.

Sprinkle a pastry board or work surface with a little extra caster sugar and very lightly roll out the mixture ¼ in (5 mm) thick. If the mixture is very soft, roll out half at a time. Cut into stars with a pastry cutter and place on buttered baking sheets. Bake for 10 to 15 minutes until the top of each star has risen a little and is light and quite dry.

Remove from the oven, allow to cool for a few minutes, then carefully transfer the biscuits to wire racks (they are still quite fragile at this stage). Leave to cool.

Meanwhile make the icing by beating the egg whites with the icing sugar and lemon juice until very thick and white.

Using a pointed knife, spread a little icing on each star, making sure there is icing on each point but leaving a narrow margin round the edges.

When the icing has nearly set, sprinkle a few *nonpareils* on to the centre of each star and leave to set completely.

Store in biscuit tins when quite dry. You can thread the stars, using a thin skewer and some thin cord or ribbon, for hanging on the Christmas tree.

Pancakes

****

Pancake Day comes but once a year, but it would be a pity to forget about pancakes all the other months, for they can be served at any meal, and the variety of their fillings, sweet and savoury, can be all but infinite. They are easily made in large batches and they freeze excellently, either filled or plain.

To freeze unfilled, stack them in convenient batches – 6, 8, 12 or more according to the numbers you are likely to need later on. Wrap closely in clingfilm or foil, label with the number of pancakes in each packet and freeze. When you come to use them, they will separate quite easily with a small, sharp-pointed knife, and if you lay them out separately will thaw within a very short time. The smaller the pancakes, the easier they are to handle. Large pancakes will probably need to be interleaved with clingfilm or greaseproof paper, so that they will separate without breaking.

If the pancakes are to be eaten cold, they can be filled as soon as they have thawed, and served straight away.

Savoury Pancake Mixture

8 oz (225 g) plain flour
1 tsp (5 ml) salt
½ tsp (2.5 ml) curry powder
  (optional)
2 whole eggs
1 egg yolk

½ pt (300 ml) milk
½ pt (300 ml) water or ¼ pt (150 ml)
  water and ¼ pt (150 ml) beer
dash of soda water (optional)
oil or lard for frying

Sift the flour with the salt and curry powder, if you are using it, into a mixing bowl, make a well in the centre and pour in the whole eggs and the egg yolk. Using a wooden spoon, gradually draw the flour into the centre, slowly adding the milk or the water and beer and beating well to make a smooth batter. Set aside for about 1 hour.

You can also make pancake batter in a blender or food processor. Reverse the process by first blending or processing the eggs and liquids together, then add the sifted flour, salt and curry powder. Set aside as above.

When you are ready to make the pancakes, add the soda water, if using, to the batter. Put a very small piece of lard or a very little oil into a heavy frying pan, wiping off the excess, and when it is really hot pour in a little of the batter – just enough to cover the bottom of the frying pan. Cook the pancake on both sides over a moderate heat. Repeat until all the batter is used up.

With a well-seasoned pan, you will probably need to regrease the pan only for every second or third pancake.

The quantities given here will make 15 to 18 pancakes of 9-in (23-cm) diameter, or about 30 pancakes of 5-in (12-cm) diameter. They should be as thin as possible.

Pancakes with Smoked Salmon or Caviar and Sour Cream Filling

Serves 6

This makes an elegant and deceptively economical start to a dinner.

6 small pancakes (see p. 139)
¼ pt (150 ml) sour cream
2–4 oz (50–100 g) smoked salmon trimmings or a 1–2 oz (25–50 g) jar caviar or lumpfish roe

salt and freshly ground pepper

a little finely chopped parsley or chives

Frozen pancakes must be allowed to thaw thoroughly, as they are not heated for this dish.

Whip the sour cream lightly. Fold in the finely chopped smoked salmon or caviar and the seasoning.

Spoon a portion of the filling along the middle of each pancake and roll up. Sprinkle with the chopped parsley or chives before serving.

Pancakes with Chicken or Beef Filling

Serves 6 as a starter or 4 as a main course

An excellent way to stretch a small amount of cooked meat.

12 small or 6–8 medium pancakes
  (see p. 139)
1 oz (25 g) butter
1 oz (25 g) plain flour
¼ pt (150 ml) dry white wine
¼ pt (150 ml) chicken or beef stock
¼ pt (150 ml) double or sour cream

1 tsp (5 ml) made mustard
salt and freshly ground pepper
8 oz (225 g) finely diced cooked
  chicken or beef

*1 tbls (15 ml) grated parmesan or
  cheddar cheese*

Frozen pancakes must be allowed to thaw thoroughly.

Melt the butter in a heavy saucepan, add the flour and cook for 2 to 3 minutes, without allowing to brown. Add the wine, stir until smooth and cook for a minute before adding the stock, and then the cream.

Stir in the mustard and season to taste. Fold in the meat and heat through.

Spoon a portion of the filling along the middle of each pancake and roll up. Sprinkle with a little grated cheese and brown under the grill before serving.

Crespolini (Pancakes with Spinach, Mushroom and Ricotta Filling)

Serves 6 as a starter, 4 as a main dish

12 small or 6–8 medium pancakes
  (see p. 139)

Filling
6 oz (175 g) cooked or frozen spinach
  (about 12 oz (350 g) fresh)
6 oz (175 g) mushrooms
1 oz (25 g) butter

6 oz (175 g) ricotta cheese
1 egg
salt and freshly ground pepper
freshly grated nutmeg

Sauce
1½ oz (40 g) butter
1½ oz (40 g) plain flour
1 scant pt (500 ml) milk
4 oz (100 g) grated cheddar cheese

salt and freshly ground pepper

*2 oz (50 g) grated parmesan or cheddar
  cheese*

Frozen pancakes must be allowed to thaw thoroughly.

First make the filling. If you are using fresh spinach, wash it well and cook gently without water until it is tender. Drain well. If you are using frozen spinach, warm the solid block through over the lowest possible heat and cook until tender. Mash the spinach finely or put through a coarse mouli.

Cook the sliced mushrooms gently in the butter for about 5 minutes. Put through a coarse mouli.

Mix well together the spinach, mushrooms, ricotta cheese and beaten egg. Season with salt and generously with freshly ground pepper and grated nutmeg.

Spoon a portion of the filling along the middle of each pancake and roll up. Arrange in a single layer in a shallow ovenproof dish.

To make the sauce, melt the butter in a saucepan and stir in the flour. Allow to cook gently, stirring, for 2 to 3 minutes. Slowly add the milk, stirring, until the sauce is thick and smooth. Stir in the grated cheese and season.

Pour the sauce evenly over the pancakes.

To serve immediately: sprinkle with the grated parmesan or cheddar cheese and bake in the centre of a moderately hot oven (200°C, 400°F, gas 6) for 30 to 40 minutes, until hot and bubbling.

To freeze: allow to cool, then wrap well and freeze.

To serve after freezing: either allow to thaw and reheat as above; or transfer straight from the freezer (having first sprinkled on the cheese) to a hot oven (220°C, 425°F, gas 7) and bake for about 30 minutes, then lower the oven to 180°C, 350°F, gas 4 for a further 45 minutes or so. Cover with a piece of foil if the pancakes show signs of overbrowning.

Pancakes with Ham and Almond Filling

Serves 6 as a starter, 4 as a main dish

A good way of using up the knuckle of a cooked ham or any other leftover ham.

12 small or 6–8 medium pancakes
(see p. 139)

Filling

3 oz (75 g) blanched almonds
6 oz (175 g) cooked ham
¾ oz (20 g) butter
¾ oz (20 g) plain flour
⅓ pt (200 ml) half dry white wine and
half stock

1½ oz (40 g) grated cheddar cheese
1–2 tbls double or whipping cream
freshly ground pepper

Sauce

1½ oz (40 g) butter
1½ oz (40 g) plain flour
1 scant pt (500 ml) milk
3 oz (75 g) grated cheddar cheese

freshly ground pepper

1½ oz (40 g) grated cheddar cheese

Frozen pancakes must be allowed to thaw thoroughly.

First make the filling. Chop finely the almonds and put under a low grill or in a very gentle oven until they are light brown. Do not allow them to get dark.

Mince the ham.

Melt the butter in a small saucepan, stir in the flour and allow to cook, stirring, for 2 to 3 minutes. Slowly add the white wine and stock, stirring, until the sauce is smooth and very thick. Stir in the ham, almonds, grated cheese and cream and season with a little pepper, but do not add salt as the ham will be salty.

Spoon a portion of the filling along the middle of each pancake and roll up. Arrange in a single layer in a shallow ovenproof dish.

To make the sauce, melt the butter in a saucepan and stir in the flour. Allow to cook gently, stirring, for 2 to 3 minutes. Slowly add the milk, stirring, until the sauce is thick and smooth. Stir in the grated cheese and season with a little pepper.

Pour the sauce evenly over the pancakes.

To serve immediately: sprinkle the cheese over the pancakes and bake in a moderately hot oven (200°C, 400°F, gas 6) for about 30 minutes, until hot and bubbling.

To freeze: allow to cool, then wrap and freeze.

To serve after freezing: either allow to thaw and reheat as above; or transfer straight from the freezer (having first sprinkled on the cheese) to a hot oven (220°C, 425°F, gas 7) and bake for about 30 minutes, then lower the oven to 180°C, 350°F, gas 4 for a further 45 minutes or so. Cover with a piece of foil if the pancakes show signs of overbrowning.

Pancakes with Ham and Pineapple Filling

Serves 6 as a starter, 4 as a main course

12 small or 6–8 medium pancakes (see p. 139)
1½ oz (40 g) butter
1½ oz (40 g) plain flour
1 scant pt (500 ml) milk

salt and freshly ground pepper
8 oz (225 g) thinly sliced ham
1 × 15-oz (425-g) can pineapple slices or pieces

3 oz (75 g) grated cheese

Frozen pancakes must be allowed to thaw thoroughly.

Heat the oven to 200°C, 400°F, gas 6.

Melt the butter in a saucepan and stir in the flour. Allow to cook gently, stirring, for 2 to 3 minutes. Slowly add the milk, stirring until the sauce is smooth and thick. Season.

Lay enough ham over each pancake to cover. Drain the pineapple and cut into small pieces. Divide among the pancakes, laying them over the middle of the ham. Roll up and arrange in a shallow ovenproof dish in one layer.

Pour the sauce over the pancakes and sprinkle the grated cheese on top. Heat through in the oven for 30 to 40 minutes, until bubbling. Finish by putting under the grill for about 5 minutes, so that the cheese becomes a rich brown.

Pancakes with Shrimp Filling

Serves 6 as a starter, 4 as a main dish

12 small or 6–8 medium pancakes
   (see p. 139)

Filling

12 oz (350 g) frozen shrimps
1 medium onion
3 oz (75 g) butter
2 tbls (30 ml) dry white wine
1½ oz (40 g) plain flour
½ pt (300 ml) milk

¼ pt (150 ml) double or
   whipping cream
1 egg yolk
1 tbls (15 ml) chopped parsley
salt and freshly ground pepper

Sauce

1½ oz (40 g) butter
1½ oz (40 g) plain flour
1 scant pt (500 ml) milk
4 oz (100 g) grated cheddar cheese

salt and freshly ground pepper

*2 oz (50 g) grated parmesan or cheddar
   cheese*

Frozen pancakes and shrimps must be allowed to thaw thoroughly.

Heat the oven to 200°C, 400°F, gas 6.

Cook the finely chopped onion gently in half the butter until soft. Add the thoroughly drained shrimps. Cook gently, stirring, for 3 minutes. Add the wine, turn up the heat, and cook until the wine has almost evaporated.

Melt the remaining butter in another saucepan, add the flour and stir over a low heat for 2 to 3 minutes. Slowly add the milk, stirring until the mixture is thick and smooth. Beat in the cream and the egg yolk, but do not allow to boil. Add the shrimp mixture and the parsley and season to taste.

Spoon a portion of the filling along the middle of each pancake and roll up. Arrange in a single layer in a shallow ovenproof dish.

To make the sauce, melt the butter in a saucepan and stir in the flour. Allow to cook gently, stirring, for 2 to 3 minutes. Slowly add the milk, stirring, until the sauce is thick and smooth. Stir in the grated cheese and season.

Pour the sauce evenly over the pancakes. Sprinkle with grated parmesan or cheddar cheese and bake in the centre of the oven for 30 to 40 minutes, until hot and bubbling.

Pancakes with Chicken and Mushroom Filling

Serves 6 as a starter, 4 as a main course

12 small or 6–8 medium pancakes
  (see p. 139)

Filling

1 large onion
1 oz (25 g) butter or chicken fat
6 oz (175 g) mushrooms
6 oz (175 g) cooked chicken

1–2 tbls (15–30 ml) parsley (optional)
a little of the sauce (see below)
salt and freshly ground pepper

Sauce

2 oz (50 g) butter
2 oz (50 g) plain flour
1¼ pts (750 ml) milk or half milk and
  half chicken stock
4 oz (100 g) grated parmesan or
  cheddar cheese

salt and freshly ground pepper
freshly grated nutmeg

*2 oz (50 g) grated parmesan or cheddar
  cheese*

Frozen pancakes must be allowed to thaw thoroughly.

Heat the oven to 200°C, 400°F, gas 6.

First make the filling. Cook the finely chopped onion gently in the butter or chicken fat until soft. Add the roughly chopped mushrooms and cover. Cook gently for about 15 minutes.

Remove from the heat and either put through a coarse mouli, together with the chicken, or chop both finely. Stir in the parsley, if you are using it. Mix in a little of the sauce (see below), so that the filling is moist but not runny. Season.

Make the sauce while the filling is cooking. Melt the butter in a saucepan and stir in the flour. Allow to cook gently, stirring, for 2 to 3 minutes. Gradually stir in the milk, or milk and chicken stock, until the sauce is thick and smooth. At this stage use a little of the sauce to moisten the filling (see above). Stir the grated cheese into the remaining sauce and season with salt and pepper and plenty of freshly grated nutmeg.

Spoon a portion of the filling along the middle of each pancake and roll up. Arrange in a single layer in a shallow ovenproof dish and pour the sauce evenly over the top.

To serve immediately: sprinkle with the grated parmesan or cheddar cheese and heat through in the oven for about 30 minutes, until hot and bubbling.

To freeze: allow to cool, then wrap well and freeze.

To serve after freezing: either allow to thaw and reheat as above; or transfer straight from the freezer (having first sprinkled on the cheese) to a hot oven (220°C, 425°F, gas 7) for about 30 minutes, then lower the oven to 180°C, 350°F, gas 4 for a further 45 minutes or so. Cover with a piece of foil if the pancakes show signs of overbrowning.

Pancakes with Stilton Filling

Serves 6 as a starter

These tasty pancakes are best eaten as a starter.

12 small or 6–8 medium pancakes (see p. 139)
1½ oz (40 g) butter
1½ oz (40 g) plain flour
scant ¾ pt (400 ml) milk
2–3 tbls (30–45 ml) double cream

3 oz (75 g) stilton
freshly ground pepper

1½–2 oz (40–50 g) grated parmesan cheese

Frozen pancakes must be allowed to thaw thoroughly.

Heat the oven to 180°C, 350°F, gas 4.

Melt the butter in a saucepan, add the flour, and stir over a low heat for 2 to 3 minutes. Gradually add the milk, stirring, until the sauce is thick and smooth. Add the cream and the crumbled stilton, stirring off the heat until the cheese has melted. Season with pepper.

Spread a portion of the filling evenly over each pancake and roll up. Arrange in a single layer in a generously buttered ovenproof dish. Sprinkle over the grated parmesan cheese and heat through for about 20 minutes, until the pancakes are golden and bubbling.

Note: pancakes can also be filled with any of the following mixtures:
Mussels au Gratin (p. 48)
Prawns in Cream and Brandy Sauce (p. 50)
The filling for Prawn and Mushroom Vol-au-Vents (p. 52)
Scallops au Gratin (p. 55).

Sweet Pancake Mixture

These quantities will make 15–18 pancakes of 9-in (23-cm) diameter, or about 30 pancakes of 5-in (12-cm) diameter.

8 oz (225 g) plain flour
pinch of salt
1 tbls (15 ml) icing sugar
2 whole eggs
2 egg yolks

1 pt (600 ml) milk
1 tbls (15 ml) brandy
1 oz (25 g) melted butter
dash of soda water

Mix the batter as for savoury pancakes (see p. 139). Set aside for about 1 hour, then whisk in the melted butter and the soda water just before you begin to cook the pancakes, as described on p. 140. Make them as thin as possible.

These pancakes are delicious served hot with lemon juice and plenty of sugar, syrup, maple syrup or honey, or with any of the following fillings.

SWEET FILLINGS

For these recipes the pancakes should be heated through before filling. The quantities given below are enough to make a dessert for 6 people, allowing one large or two small pancakes per person.

Praline Butter

2 oz (50 g) sugar
1 tbls (15 ml) water
2 oz (50 g) blanched whole almonds
2 oz (50 g) butter

1 oz (25 g) caster sugar
1 tbls (15 ml) rum (optional)

1 tbls (15 ml) icing sugar

Dissolve the sugar in the water in a small, heavy saucepan. When it begins to turn golden, add the almonds and stir until they are all covered with golden brown toffee.

Pour on to an oiled plate or board and leave to cool. Crush to a fine powder with a rolling pin or in a blender.

Cream the butter with the caster sugar until light and fluffy, then beat in the praline and the rum, if you are using it. Spread the warm pancakes with the mixture and roll up. Arrange them on a serving dish and dust with the sifted icing sugar before serving.

Coffee Marron

1 tbls (15 ml) strong black coffee
2 tbls (30 ml) rum or brandy
8 oz (225 g) sweetened chestnut
  purée

½ pt (300 ml) double cream

1 tbls (15 ml) icing sugar

Blend the coffee and half the rum or brandy into the chestnut purée. Whip the cream with the remaining rum or brandy and fold half of this into the chestnut purée. Spread the warm pancakes with the mixture and roll up. Arrange them on a serving dish and cover with the remaining cream. Dust with a little sifted icing sugar before serving.

Cream Cheese and Almond

2 oz (50 g) flaked almonds
8 oz (225 g) light cream cheese
2 oz (50 g) sugar

2 tbls (30 ml) sweet sherry

1 tbls (15 ml) icing sugar

Toast or fry the almonds until golden-brown.

Beat the cream cheese with the sugar and half the sherry until smooth. Spread the warm pancakes with the mixture, sprinkle with half the almonds and roll up.

Arrange the pancakes in a serving dish and sprinkle with the remaining sherry and almonds and the sifted icing sugar before serving.

Apple Filling

1 lb (500 g) eating apples
2 oz (50 g) butter
squeeze of lemon juice
sugar to taste
2 tbls (30 ml) double cream

2 tbls (30 ml) calvados, brandy or
  rum (optional)

1 tbls (15 ml) caster sugar

Peel and core the apples and slice them thinly.

Melt the butter in a saucepan, add the apples and lemon juice, cover and cook over a gentle heat until soft.

Add sugar to taste and allow to cool slightly.

Blend in the cream and calvados, brandy or rum, if you are using it. Spoon a portion of the filling along the centre of each hot pancake and roll up. Sprinkle with caster sugar, put under a hot grill until the sugar is melted and serve at once.

Crêpes Suzette

8–10 small and very thin sweet
   pancakes (see p. 147)
4 oz (100 g) butter
4 oz (100 g) sugar
grated rind and juice of 1 lemon
grated rind and juice of 2 oranges

1 tbls (15 ml) icing sugar
2 tbls (30 ml) orange liqueur (Grand
   Marnier or orange curaçao)
2 tbls (30 ml) brandy

Frozen pancakes must be allowed to thaw thoroughly.

Melt the butter and sugar in a large frying pan. When it is bubbling hot, add the lemon and orange rind and juice.

Dip each pancake briefly in this sauce, then fold into four and arrange on a heated serving dish. Sprinkle with sifted icing sugar and keep warm.

Add the liqueur and brandy to the sauce in the pan, heat quickly and set alight. Pour the sauce over the pancakes and bring flaming to the table.

You can if you prefer add the liqueur and brandy to the sauce at the table and set it alight there.

Yeast and Bread Baking

❊❊❊❊❊❊❊❊❊❊❊❊❊❊❊❊❊❊

Yeast has a bad reputation for being temperamental, and rarely gets sufficient credit for its adaptability and willingness to stop and start working at the cook's behest.

Buy fresh (compressed) yeast in 8-oz (225-g) or 1-lb (500-g) blocks. In this way, provided you get it into the freezer quickly, you will probably have fresher yeast than ever before, for once it has been crumbled off in the shop it immediately loses some of its freshness. Divide the yeast into ½-oz (15-g) and 1-oz (25-g) pieces. Wrap each piece in clingfilm and store all together in a polythene bag or a plastic box in the freezer. Stored in this way, the yeast will keep perfectly well for months.

When you take yeast from the freezer, give it sufficient time to come back to life before using. However, if you really need it quickly you can put it in a cup and cover with tepid (not hot) water for about 15 minutes (deduct the liquid you use from the quantity given in the recipe).

Utterly satisfying though it is to make your own bread, few people have time to indulge in the pleasure often. So it is well worth making several loaves at once, for as Elizabeth David has said, 'The freezer is surely the best bread bin to date.'

Freshly baked bread should be allowed to cool thoroughly, then placed in a polythene bag, or wrapped in clingfilm and covered with foil. It should then be frozen immediately.

To thaw, unwrap and leave at room temperature for several hours, or, if you are in a hurry or like to fill your house with the evocative smell of newly baked bread, put it straight from the freezer into a moderate oven (190°C, 375°F, gas 5) and bake for 30 to 40 minutes, depending on the size of the loaf. Bread which has been reheated in this way should be eaten the same day, as it will go stale quite quickly.

You can freeze bread dough, but in this case double the quantity of yeast. Prepare the dough, give it its first rising, knock down, and shape or place in tins. Lay a sheet of clingfilm over the top and put into large polythene bags. Store in the freezer.

When ready to cook, take out of the freezer, remove the clingfilm, and leave the loaf loosely tied in the polythene bag. Leave to thaw at room temperature for 4 to 6 hours, then put in a warm place to rise. When the dough is ready, bake as usual.

Dried yeast can be used instead of fresh, though some cooks feel that it makes the bread taste of yeast – perhaps because too much is used. But it is a valuable stand-by when there is no fresh yeast available. If you use dried yeast, you will need only about half the quantity you would need of fresh.

Wholemeal Bread

Makes 4 × 1-lb (450-g) or 2 × 2-lb (900-g) loaves

3 lb (1.5 kg) 100 per cent wholemeal flour
4 tsp (20 ml) salt
1 oz (25 g) lard
1 oz (25 g) fresh yeast or 1 rounded tbls (17.5 ml) dried yeast
1 level tbls (15 ml) barbados sugar
about 1½ pts (900 ml) tepid water
a little strong white flour

Sift the flour with the salt into a mixing bowl, rub in the lard and stand in a warm place for a few minutes.

If using fresh yeast, cream the yeast and sugar in a small bowl until liquid (this takes only a minute or two). Add a little of the tepid water and stir well together. This mixture can be used at once – there is no need to wait for it to get frothy.

If using dried yeast, dissolve the sugar in ¼ pt (150 ml) of the tepid water. Add the yeast and whisk. Stand in a warm place until frothy (about 10 minutes).

Make a well in the centre of the flour, whisk the yeast mixture for a few seconds and stir it into the flour with your fingers or a wooden spoon. Gradually add enough water to absorb all the flour. Remove to a floured board and knead for 10 minutes, adding a little of the strong white flour from time to time as the dough becomes too sticky to knead easily.

Put the dough into a greased mixing bowl, cover, and leave in a warm place to rise for 1 to 2 hours until about doubled in size.

Knead again lightly and divide into two or four, according to whether you are making 2-lb (900-g) or 1-lb (450-g) loaves. Put into the tins, which should be warm and lightly greased.

Cover and leave again in a warm place to rise (30 minutes to 1 hour). Heat the oven to 220°C, 425°F, gas 7.

When the loaves have risen nearly to the top of the tins, place them in the centre of the oven and bake for 30 to 40 minutes, until they are well-browned, have risen well above the tins, and will slide out easily. Turn out on to a wire rack and leave to cool.

To freeze: wrap, then freeze.

To serve after freezing: see p. 151.

Wholemeal Rolls

Makes about 18 rolls

Rolls are best made with all milk, or a mixture of milk and water, since if they are made with water alone the outside tends to be hard. It is useful to have a supply of them in the freezer, for they can be warmed up very quickly if you run out of bread, have unexpected visitors, or just need one or two rather than a whole loaf.

14 oz (400 g) wholemeal flour
½ tsp (2.5 ml) salt
1 oz (25 g) lard
½ oz (15 g) fresh yeast or 2 tsp
(10 ml) dried yeast

½ oz (15 g) sugar
about ⅜ pt (250 ml) milk, or
half milk and half water

Mix together the flour and salt and rub in the lard. Leave in a warm place for about 15 minutes.

If using fresh yeast, cream the yeast and sugar in a small bowl until liquid (this takes only a minute or two). Add a little of the tepid milk, or milk and water, and stir well together. This mixture can be used at once – there is no need to wait for it to get frothy.

If using dried yeast, dissolve the sugar in ¼ pt (150 ml) of the tepid milk, or milk and water. Add the yeast and whisk. Stand in a warm place until frothy (about 10 minutes).

Make a well in the centre of the flour and add the yeast mixture. Mix in enough milk, or milk and water, to make a stiff dough. Transfer to a floured board and knead for about 5 minutes, adding a little more flour from time to time if the mixture becomes too sticky.

Put the dough into a greased mixing bowl, cover, and leave to rise in a warm place (about 1 hour). Knock down and knead again briefly. Form into about 18 rolls. Place on a greased baking sheet, cover, and leave to rise in a warm place.

Heat the oven to 200°C, 400°F, gas 6.

When the rolls have risen, place in the centre of the oven and bake for 10 to 15 minutes. Leave to cool on a wire rack.

To freeze: freeze in a polythene bag.

To serve after freezing: if you want to serve the rolls hot, transfer them straight from the freezer to a moderate oven (180°C, 350°F, gas 4) for about 10 minutes. Otherwise leave to thaw at room temperature for 1 to 2 hours.

Grant Loaf

Makes 4 × 1-lb (450-g) or 2 × 2-lb (900-g) loaves

The beauty of this wholemeal loaf is that it needs no kneading. It has a closer texture than ordinary wholemeal bread, and you can really taste the goodness.

3 lb (1.5 kg) 100 per cent wholemeal flour
1 tbls (15 ml) salt
1 oz (25 g) fresh yeast or 1 rounded tbls (17.5 ml) dried yeast
1 tbls (15 ml) barbados sugar
2 pts (1.2 L) tepid water

Sift the flour with the salt into a mixing bowl and leave in a warm place for about 15 minutes.

If using fresh yeast, cream the yeast and sugar in a small bowl until liquid (this takes only a minute or two). Add a little of the tepid water and stir well together. The mixture can be used at once – no need to wait for it to get frothy.

If using dried yeast, dissolve the sugar in ¼ pt (150 ml) of the tepid water. Add the yeast and whisk. Stand in a warm place until frothy (about 10 minutes).

Make a well in the centre of the flour, whisk the yeast mixture for a few seconds, and stir into the flour. Gradually stir in the remaining water, and mix thoroughly to a rather soft dough. Divide between the warm, lightly-greased bread tins. Cover and leave in a warm place for the loaves to rise (1 to 2 hours).

Heat the oven to 190°C, 375°F, gas 5.

When the loaves have risen to the top of the tins, place them in the centre of the oven and bake for 35 to 45 minutes, until they are well-browned, have risen well above the tins, and will slide out easily. Turn out on to a wire rack and leave to cool.

To freeze: wrap, then freeze.

To serve after freezing: see p. 151.

White Bread

Makes 5 × 1-lb (450-g) or 2 × 2-lb (900-g) and 1 × 1-lb (450-g) loaves

3 lb (1.5 kg) strong white flour	1 tsp (5 ml) sugar
1 oz (25 g) salt	1½ pts (900 ml) warm water
1 oz (25 g) fresh yeast or 1 rounded	2 oz (50 g) lard or butter
tbls (17.5 ml) dried yeast	a little extra flour

Sift the flour with the salt into a mixing bowl and stand in a warm place for a few minutes.

If using fresh yeast, cream the yeast and sugar in a small bowl until liquid (this takes only a minute or two). Add a little of the tepid water and stir well together. This mixture can be used at once – no need to wait for it to get frothy.

If using dried yeast, dissolve the sugar in ¼ pt (150 ml) of the tepid water. Add the yeast and whisk. Stand in a warm place until frothy (about 10 minutes).

Rub the lard or butter into the flour. Make a well in the centre, pour in the yeast mixture and stir it into the flour with your fingers or a wooden spoon. Gradually add the remaining water until all the flour has been absorbed. Knead for 10 minutes, adding a little extra strong white flour from time to time as the dough becomes too sticky to knead easily.

Put into a greased mixing bowl, cover, and leave in a warm place for the dough to rise until it has about doubled in size. The time this will take depends on the temperature – it may take anything from 1½ to 2 hours, or even longer.

Knock back the dough and turn on to a lightly floured board. Divide into 1-lb (450-g) or 2-lb (900-g) pieces. Knead each piece and put into lightly greased tins. Cover and leave in a warm place to rise again (35 to 50 minutes).

Heat the oven to 230°C, 450°F, gas 8.

Place the loaves in the centre of the oven and bake for 15 minutes. Turn the oven down to 200°C, 400°F, gas 6, and bake for a further 10 to 15 minutes for smaller loaves and 20 to 30 minutes for larger ones.

Turn out of the tins and leave on a rack to cool.

To freeze: wrap well, then freeze.

To serve after freezing: see p. 151.

Irish Soda Bread

Makes 4 small loaves

2 lb (900 g) wholemeal flour	2 tsp (10 ml) salt
1 rounded tbls (17.5 ml) bicarbonate of soda	about 1 pt (600 ml) buttermilk or sour milk

Heat the oven to 200°C, 400°F, gas 6.

Sift the flour with the bicarbonate of soda and salt into a mixing bowl and mix well. Stir in the buttermilk or sour milk, adding a little extra liquid if necessary to make a soft, but not wet, dough. Place the dough on a floured board and flatten into a circle about 1½ in (3.5 cm) thick. Transfer to a baking sheet and make a large cross on the top with a floured knife.

Bake in the centre of the oven for 30 to 35 minutes, until the bread has risen and is golden-brown and a skewer inserted into the centre comes out clean.

When the bread is quite cold cut into quarters.

To freeze: wrap well, then freeze.

To serve after freezing: allow to thaw at room temperature for about 3 hours.

Note: fresh milk can be used instead of buttermilk or sour milk, but in this case add 1 heaped tsp (7.5 ml) cream of tartar to the dry ingredients.

Wholemeal Scones

12 oz (350 g) wholemeal flour
4 oz (100 g) plain flour
½–1 tsp (2.5–5 ml) salt
1 tbls (15 ml) brown sugar
4 oz (100 g) lard or margarine

2 rounded tsp (12.5 ml) baking
   powder
about ⅓ pt (200 ml) milk, preferably
   sour

Heat the oven to 240°C, 475°F, gas 9.

Sift together the wholemeal flour, white flour, salt and sugar. Rub in the lard or margarine and stir in the baking powder. Add enough milk to make a stiff dough.

Turn on to a floured board and knead lightly. Roll out about ¾ in (2 cm) thick, and cut into small rounds with a pastry cutter.

Place on a floured baking sheet and bake for 10 to 15 minutes, until the scones are light brown. Cool on a wire rack.

To freeze: as soon as the scones are cool, put into a plastic bag and freeze.

To serve after freezing: place in a moderate oven (180°C, 350°F, gas 4) for about 15 minutes.

White Scones

1 lb (450 g) self-raising flour
½–1 tsp (2.5–5 ml) salt
1 tbls (15 ml) brown sugar
2 tsp (10 ml) bicarbonate of soda

2 tsp (10 ml) cream of tartar
6 oz (175 g) margarine
2 eggs
a little milk, preferably sour

Heat the oven to 240°C, 475°F, gas 9.

Sift the flour with the salt, sugar, bicarbonate of soda and cream of tartar into a mixing bowl. Rub in the margarine. Beat the eggs lightly and stir in, adding just enough milk to make a stiff dough.

Turn on to a floured board and knead lightly. Roll out about ¾ in (2 cm) thick, and cut into small rounds with a pastry cutter.

Place on a floured baking sheet and bake for 10 to 15 minutes, until the scones are golden-brown. Cool on a wire rack.

To freeze: as soon as the scones are cool, put into a plastic bag and freeze.

To serve after freezing: place in a moderate oven (180°C, 350°F, gas 4) for about 15 minutes.

Cheese Loaf

Makes 2 × 1-lb (450-g) loaves

Granary flour and cheese make a happy combination in this loaf. It does not taste perceptibly of cheese, and can be used to accompany any food for which you would use ordinary bread.

1½ lb (675 g) granary flour	approx ⅝ pt (375 ml) tepid water
2 tsp (10 ml) salt	2 tsp (10 ml) barbados sugar
½ oz (15 g) lard	4 oz (100 g) grated cheddar cheese
½ oz (15 g) fresh yeast or 2 rounded tsp (12.5 ml) dried yeast	a little strong white flour

Sift the flour with the salt into a mixing bowl, rub in the lard, and leave in a warm place for a few minutes.

If using fresh yeast, cream the yeast and sugar in a small bowl until liquid (this takes only a minute or two). Add a little of the tepid water and stir well together. This mixture can be used at once – no need to wait for it to get frothy.

If using dried yeast, dissolve the sugar in ¼ pt (150 ml) of the tepid water. Add the yeast and whisk. Stand in a warm place until frothy (about 10 minutes).

Make a well in the centre of the flour, whisk the yeast mixture for a few seconds and stir it into the flour with your fingers or a wooden spoon. Stir in the cheese. Gradually add enough water to absorb all the flour and cheese. Knead for 10 minutes, adding a little of the strong white flour from time to time as the dough becomes too sticky to knead easily.

Put into a greased mixing bowl, cover, and leave in a warm place for the dough to rise until it has about doubled in size (1 hour or longer). Knead again and divide in half. Put into warm, lightly greased tins. Cover and leave again in a warm place to rise (30 to 45 minutes).

Heat the oven to 180°C, 350°F, gas 4.

When the loaves have risen nearly to the top of the tins, place them in the centre of the oven and bake for about 30 minutes, until they are well-browned, have risen well above the tins, and will slide out easily. Turn out on to a wire rack and leave to cool.

To freeze: wrap, then freeze.

To serve after freezing: see p. 151.

Onion and Walnut Bread

Makes 2 × 1-lb (450-g) loaves

Special breads are marvellous with soup, and this loaf, served hot with plenty of good butter and a rich, thick, home-made vegetable soup, makes an excellent winter supper.

2 large onions	1 tsp (5 ml) sugar
6 tbls (90 ml) walnut or olive oil	¾ pt (450 ml) warm water
6 oz (175 g) walnuts	1½ lb (700 g) strong unbleached
1 oz (25 g) fresh yeast or 1 rounded	white flour
tbls (17.5 ml) dried yeast	2 tsp (10 ml) salt

Chop the onions fairly finely. Heat half the oil until almost smoking and fry the onions until crisp. Drain and leave to cool.

Chop the walnuts fairly finely.

If using fresh yeast, cream the yeast and sugar in a small bowl until liquid (this takes only a minute or two). Add a little of the tepid water and stir well together. This mixture can be used at once – no need to wait for it to get frothy.

If using dried yeast, dissolve the sugar in ¼ pt (150 ml) of the tepid water. Add the yeast and whisk. Stand in a warm place until frothy (about 10 minutes).

Sift the flour with the salt into a mixing bowl, make a well in the centre and add the yeast mixture and the remaining water and oil. Knead into a dough and leave in a warm place to rise for 1 to 2 hours.

Knock down and knead again, this time adding the onions and the walnuts. Work them into the dough, then divide in half and place in 2 × 1-lb (450-g) oiled loaf tins. Leave to rise again for at least 30 minutes (this mixture will not rise as much as a plain dough).

Bake in a moderate oven (190°C, 375°F, gas 5) for 50 minutes to 1 hour. Turn out the loaves, and if the sides and bottom are still a little moist place upside down on baking sheets and return to the oven for another 5 to 10 minutes. Turn out on to a wire rack and leave to cool.

To freeze: wrap, then freeze.
To serve after freezing: see p. 151.

Variations: you can make Onion and Olive Bread by substituting 4 oz (100 g) black olives, stoned weight, and coarsely chopped, for the walnuts. Use olive oil for this bread.

Croissants

Makes 12–20 croissants

For a leisurely Sunday breakfast treat, nothing can beat fresh, hot croissants with coffee. They are expensive to buy, but excellent croissants can be made in quantity at home, stored uncooked in the freezer and freshly baked when required.

1 oz (25 g) fresh yeast	7 oz (200 g) butter
4 tbls (60 ml) tepid water	1 lb (450 g) strong white flour
1 oz (25 g) sugar	2 tsp (10 ml) salt
½ pt (300 ml) milk	

Glaze

1 egg yolk	2 tbls (30 ml) milk

Cream the yeast with the water and 1 tsp (5 ml) of the sugar.

Bring the milk to the boil, then remove from the heat. Add 1 oz (25 g) of the butter and the remaining sugar, and stir until dissolved. Leave until lukewarm.

Sift the flour with the salt into a mixing bowl. Make a well in the centre, pour in the yeast mixture and the milk and butter mixture, and gradually work into the flour with a wooden spoon. Knead the mixture until it forms a smooth and springy dough and leaves the sides of the bowl clean.

Cover with a clean cloth and leave in a warm place until the dough has risen to double its bulk (1 to 2 hours).

Meanwhile, soften the remaining butter on a plate, working it with a palette knife until it is quite smooth. Leave at room temperature.

When the dough is ready, knock it down and knead on a floured surface for a few minutes; then roll into a rectangle three times as long as it is wide.

Divide the softened butter into three parts, and spread the first third over two-thirds of the dough. Fold the unbuttered third of dough over, and then fold over again to make a roughly square parcel. Seal the edges with a rolling pin, give the pastry a quarter turn to the right, and roll out again into a rectangle. Fold this over three times into a square, then leave the dough to rest in the refrigerator for 30 minutes.

Repeat this operation twice more, until the butter has been used up. Leave the dough in the refrigerator for a final rest of 1 hour.

On a floured surface, roll out the dough to a large rectangle, about ¼ in (5 mm) thick. Cut this into 4–6-in (10–15-cm) squares, depending on how large you want your croissants. Cut each square in half diagonally, to form two triangles. Roll up each triangle from the wide edge towards the point, and bring the ends forward to form a crescent shape with the central point underneath.

Beat the egg yolk with the milk for the glaze, and use to brush each croissant.

The croissants are now ready, either for immediate baking, for storing in the refrigerator overnight to be baked fresh for breakfast the next morning, or for storing in the freezer.

To serve immediately: bake in a moderately hot oven (190°C, 375°F, gas 5) for 15 to 20 minutes.

If the croissants have been kept overnight in the refrigerator, remove them 30 minutes before baking.

To freeze: open-freeze the croissants on baking sheets, then store in a large polythene bag, preferably interwrapping each croissant in freezer tissue first.

To serve after freezing: take out as many croissants as you need, and either leave in the refrigerator overnight and bake as above, or remove from the freezer and leave at room temperature for 1 hour before baking.

The croissants can also be stored in the freezer after baking. Transfer straight from the freezer to a hot oven (200°C, 400°F, gas 6) and bake for 20 to 25 minutes before serving.

Banana Bread

Makes 2 × 1-lb (450-g) loaves

This is halfway between a loaf and a cake, and can be eaten with or without butter. It also makes a pleasant accompaniment to a fruit dessert of any kind.

4 oz (100 g) butter or margarine	1 lb (450 g) wholemeal flour
8 oz (225 g) sugar	6 tbls (90 ml) sour cream or yoghurt
2 eggs	approx 4 oz (100 g) chopped walnuts
4 large or 6 small bananas	or pecans (optional)

Heat the oven to 180°C, 350°F, gas 4.

Butter 2 × 1-lb (450-g) loaf tins.

Cream the butter or margarine with the sugar until light and fluffy. Beat in the eggs and the bananas, mashed to a smooth pulp. Add the flour alternately with the sour cream or yoghurt, beating each in well until thoroughly amalgamated. Add the nuts if you are using them.

Divide the mixture between the prepared tins. Bake in the oven for about 1 hour, until the loaves feel just firm to the touch and a skewer inserted into the centre comes out clean.

Turn out on to a wire rack to cool.

To freeze: as soon as the bread is cool, wrap, then freeze.

To serve after freezing: unwrap the bread and thaw at room temperature for 3 to 4 hours.

PIZZAS

Pizzas are a marvellous stand-by, as they freeze excellently and can be reheated straight from the freezer (though they are moister if allowed to thaw at room temperature for 1 to 2 hours). The recipe for the dough mixture given below uses yeast, although pizzas can also be made without yeast (see *The Penguin Freezer Cookbook*). Pizzas made without yeast are lighter, with a slightly cakey texture, and are quick to prepare because they do not have to rise.

You can also buy pizza bases, which can be frozen and used later with whatever topping you like.

Pizza Dough

Makes 4 pizzas each 8 in (20 cm) in diameter

1 lb (450 g) strong white flour, or
  half white and half wholemeal
1 tsp (5 ml) salt
½ oz (15 g) fresh yeast or 2 tsp
  (10 ml) dried yeast

1 tsp (5 ml) sugar
about ½ pt (300 ml) tepid water
a little olive oil

Sift the flour with the salt into a mixing bowl and leave in a warm place
for a few minutes.

If using fresh yeast, mix together in a small bowl the fresh yeast, sugar
and about ¼ pt (150 ml) of the tepid water. The mixture can be used at
once.

If using dried yeast, dissolve the sugar in about ¼ pt (150 ml) of the
tepid water. Add the yeast and whisk. Stand in a warm place until frothy
(about 15 minutes).

Make a well in the centre of the flour and pour in the yeast mixture.
Gradually add enough water to make a stiff dough. Knead for 5 minutes,
adding a little more flour if necessary. Put into an oiled bowl, cover, and
leave in a warm place to rise (1 to 2 hours).

Heat the oven to 220°C, 425°F, gas 7.

Knock the dough down again, divide into quarters, and knead each
piece briefly. Spread evenly over 4 × 8-in (20-cm) oiled flan tins (ideally
non-stick). Or roll each piece of dough into a circle and slide on to
greased baking sheets. Brush with a little oil, cover with the topping (see
below), and leave in a warm place until they have puffed up slightly
(about 20 minutes). Place in the middle of the oven and bake for about
20 minutes.

To serve immediately: serve very hot.

To freeze: cool quickly, then wrap and freeze.

To serve after freezing: if possible, allow the pizzas to thaw at room
temperature for 1 to 2 hours before reheating in a moderate oven
(180°C, 350°F, gas 4) for about 15 minutes. If you reheat straight from
the freezer, cover with foil to prevent the top from getting dry, and put in
a moderate oven (180°C, 350°F, gas 4) for about 20 minutes, perhaps
drizzling a little olive oil over the top towards the end of the cooking.

Pizza Topping

Pizza toppings are very much a matter of taste, and can be varied in lots of ways. The important point is that the topping should be generous; and no pizza is complete without tomatoes (preferably fresh and ripe, though canned can be substituted), lots of cheese and a pronounced taste of herbs.

For 4 × 8-in (20-cm) pizzas

2 medium onions
12 oz (350 g) mozzarella or mature
 cheddar cheese
1½ lb (750 g) fresh tomatoes or 1 ×
 1 lb 14 oz (850 g) canned tomatoes
about 4 oz (100 g) black olives
a few mushrooms (optional)

1 can anchovy fillets (optional)
a few thin slices of ham (optional)
salt and freshly ground pepper
2–4 tsp (10–20 ml) dried mixed herbs
 or dried oregano
about 4 tbls (60 ml) olive oil

Divide the onions, very thinly sliced and separated into rings, the sliced mozzarella or grated cheddar cheese and the sliced fresh or drained canned tomatoes between the 4 pizza bases. Garnish with the stoned and halved olives, and with the thinly sliced mushrooms, the anchovies and the ham cut into strips, if you are using them. Season with a little salt and generously with pepper, and sprinkle the herbs and olive oil on top. If you are using anchovies, omit the salt, as they are quite salty enough.

Savarin

Serves 6–8

A savarin is one of the classics of French pâtisserie.

It is not difficult to make, always looks beautiful, and is a lovely dessert for a summer lunch party. It should be made in a savarin mould, or in a ring spring-form tin.

½ oz (15 g) fresh yeast
1 oz (25 g) sugar
¼ pt (150 ml) milk
8 oz (225 g) plain flour

pinch of salt
4 eggs
4 oz (100 g) butter

For the syrup

6 oz (175 g) sugar	*12–18 maraschino cherries*
3/4 pt (450 ml) water	*1/4–1/2 pt (150–300 ml) double or*
2 strips thinly pared lemon rind	*whipping cream*
4 tbls (60 ml) kirsch or rum	*1 tbls (15 ml) caster sugar*
12–18 split almonds	*1 tbls (15 ml) kirsch or rum*

Cream the yeast with 1 tsp (5 ml) of the sugar. Warm the milk slightly and stir in.

Sift the flour with the salt into a warmed mixing bowl and make a well in the centre. Pour in the yeast mixture and gradually work in the flour. Knead or beat with a wooden spoon until the dough becomes elastic and leaves the sides of the bowl clean. Cover with a clean cloth and leave in a warm place to rise to twice its original volume (about 1 hour).

Beat the eggs lightly together with the remaining sugar. Melt the butter and add. Slowly beat this mixture into the risen dough and continue to beat with a wooden spoon until smooth and glossy. Turn into a buttered 10-in (25-cm) savarin mould or ring spring-form tin and cover again with a cloth. Leave to rise again in a warm place until the mixture comes to the top of the mould (about 30 minutes).

Heat the oven to 190°C, 375°F, gas 5. Bake the savarin until the top is lightly browned and firm to the touch, and it has shrunk away slightly from the sides of the mould. Turn out on to a wire rack.

To serve immediately: make the syrup by boiling the sugar with the water and lemon rind for 5 minutes. Leave to cool, then remove the lemon peel and add the kirsch or rum.

Place the savarin, which should still be slightly warm, on a wire rack over a large dish. Gently spoon the syrup over the savarin until it has absorbed all the syrup and is light and spongy, but has not lost its shape.

Transfer carefully to a serving dish, and decorate with the almonds and cherries.

Whip the cream lightly and fold in the sugar and kirsch or rum, and either spoon into the centre of the savarin or serve separately.

To freeze: when quite cold, place in a tin or plastic box, or replace in the baking tin. Wrap and freeze.

To serve after freezing: the savarin can be prepared for serving straight from the freezer. Place on a wire rack over a large dish and pour over half the hot syrup, but before adding to it the kirsch or rum. Continue to pour over the drained syrup until it has all been absorbed – the savarin will be fully thawed by then. Add the kirsch or rum to the remaining syrup and finish as above.

Rum Babas

Serves 6–8

If you do not have the special little round or cylindrical baba pans, you can use individual Yorkshire pudding or castle pudding moulds or patty tins. The number of babas this quantity makes will vary according to the size of your pans or moulds.

½ oz (15 g) fresh yeast	pinch of salt
1 oz (25 g) caster sugar	4 oz (100 g) butter
¼ pt (150 ml) milk	4 eggs
8 oz (225 g) plain flour	

For the syrup

6 oz (175 g) sugar	4 tbls (60 ml) rum
¾ pt (450 ml) water	¼ pt (150 ml) double or whipping cream
2 strips of thinly pared lemon rind	glacé cherries

Cream the yeast with half the sugar. Warm the milk slightly and stir it in.

Sift the flour with the salt into a warmed mixing bowl and make a well in the centre. Pour in the yeast mixture and work the flour into the liquid with a wooden spoon. Knead or beat with the wooden spoon for 5 minutes, until the dough becomes elastic and leaves the sides of the bowl clean. Cover with a cloth and leave to rise in a warm place until the dough has nearly doubled (about 1 hour).

Butter some baba pans or small individual moulds.

Melt the remaining butter gently.

Beat the eggs with the remaining sugar and pour in the butter.

Beat this mixture slowly into the dough and continue to beat until smooth and glossy. Half-fill each mould and leave to rise again in a warm place until the dough has risen to the tops of the moulds (about 30 minutes).

Heat the oven to 220°C, 425°F, gas 7. Bake the babas in the centre of the oven until the tops are lightly browned and quite firm to the touch, and they have shrunk slightly from the sides of the pans. Leave to cool a little, then turn out on to a wire rack.

To serve immediately: make the syrup by dissolving the sugar in the water in a saucepan, add the lemon rind and bring to the boil. Boil for 5 minutes. Leave to cool, then remove the lemon rind from the syrup and add 3 tbls (45 ml) of the rum.

Place the babas on a large dish and gently spoon the syrup over each one. The babas should be allowed to absorb as much syrup as they can without losing their shape. Put them back on the wire rack, placed over a dish, so that they can drain a little, but pour the drained syrup back over them from time to time. Leave until completely cold.

Whip the cream, fold in the remaining rum and decorate each baba with a whirl of cream and a glacé cherry.

To freeze: when quite cold, pack in airtight tins or plastic boxes, then freeze.

To serve after freezing: the babas can be prepared for serving straight from the freezer. Place on a wire rack and pour over half the hot syrup but before you have added the rum. Continue to pour over the drained syrup until it has all been absorbed – the babas will be fully thawed by then. Add the rum to the remaining syrup and finish as above.

Stolle

Makes 2 large or 3 medium loaves

Halfway between a fruit loaf and a fruit cake, a Stolle is traditionally eaten in Germany at Christmas and Easter. It is much lighter than our Christmas cake, and can be eaten for breakfast, with morning coffee or at tea-time, either plain or buttered.

1 lb (450 g) currants
10 oz (275 g) sultanas
4 tbls (60 ml) rum
2¼ lb (1 kg) plain flour
2 oz (50 g) yeast
½ pt (300 ml) milk
4 oz (100 g) sugar
14 oz (400 g) butter

6 oz (175 g) chopped mixed peel
4 oz (100 g) chopped blanched
   almonds

melted butter
icing sugar

Soak the currants and sultanas in the rum.

Sift the flour into a warmed mixing bowl, make a well in the centre and crumble in the yeast.

Warm the milk to bloodheat, add a spoonful of the sugar and then stir the milk into the yeast. Work a little of the flour into the yeast mixture, then sprinkle with a little more flour, cover with a tea towel and leave in a warm place to rise (about 15 minutes).

When the yeast mixture has trebled in volume – it will rise up like a little ball in the middle of the flour – melt the butter gently and work the flour and all the remaining ingredients into a dough, leaving the fruit and nuts to the last.

Knead the dough well, then put back into the bowl and leave to rise for about 1 hour, or until it has doubled in volume. Knock down and leave to rise again.

When the dough has once more doubled in volume (about 30–45 minutes), divide into two or three and roll out each one into an oblong about 1½ in (3.5 cm) thick. Fold one side towards the centre lengthwise, to form a split loaf shape.

Place on buttered baking sheets and leave to rise again for about 15 minutes, then bake in a moderate oven (180°C, 350°F, gas 4) for about 1 hour, until the loaves have risen and are firm, and have a light brown crust. Remove from the oven and leave to cool slightly in the tins, then turn out on to wire racks.

To serve immediately: when the loaves are cool, brush with a little melted butter and dust thickly with sifted icing sugar.

To freeze: wrap the loaves in clingfilm or foil as soon as they have cooled and store in the freezer.

To serve after freezing: leave to thaw at room temperature for 6 to 8 hours, or preferably overnight. Finish as above.

Pastry

✳✳✳✳✳✳✳✳✳✳✳✳✳✳✳✳✳✳✳✳

In the first *Penguin Freezer Cookbook* we gave methods for making different kinds of shortcrust pastry, and recipes for using it. In the present book there are some additional recipes, as well as sections on choux and fila pastry.

SHORTCRUST PASTRY

Not only does this freeze well, baked or raw – freezing seems actually to improve the consistency. The dough can be frozen raw, divided into the quantities in which you are likely to use it, but it takes a long time to thaw and never seems to be as light. So it is better to roll it out and line flan tins before freezing. The pastry can then be put straight from freezer into oven.

Cornish Pasties

Makes 8 pasties

Home-made Cornish pasties are delicious either hot or cold, and marvellous for picnics. They freeze excellently, either cooked or uncooked. It may take a little practice to get them just right, so don't be discouraged if your first attempts are not quite perfect.

1 lb (500 g) braising steak
about 1 lb (500 g) potatoes
about 12 oz (350 g) onions
1½ lb (675 g) shortcrust pastry

salt and freshly ground pepper
a little butter

beaten egg and milk

Cut up the meat as finely as possible – do not mince it. Peel the potatoes and slice very thinly. Chop the onions very finely.

Roll the pastry into 8 rounds the size of a dessert plate (about 8½ in (21 cm) in diameter). This is most easily done by putting a plate upside down over the rolled pastry and carefully cutting round it with a sharp knife. Lift up one half of the pastry round and slide the rolling pin underneath – this makes it easier to fill the pasty. On the other half arrange first a layer of potatoes, then one of meat and finally one of onions. Fill the pasty as full as you possibly can, but be careful that there are no cracks or holes which would allow the filling to escape during the baking. Season very generously, and dab with two or three pieces of butter.

Moisten the edge of the half you have filled, and, using the rolling pin, flop the other half over the top. Secure the edges firmly together and make a hole in the top.

To serve immediately: brush with beaten egg and milk. Place near the top of a hot oven (230°C, 450°F, gas 8), and bake until brown (about 20 minutes). Lower the heat to 100°C, 200°F, gas ¼ and bake for a further 25 to 30 minutes. Cover with foil if the pastry is getting too brown.

To freeze uncooked: open-freeze until quite frozen, then wrap well and replace in the freezer.

To serve after freezing uncooked: brush over with egg and milk and transfer straight from the freezer to a very hot oven (240°C, 475°F, gas 9) for about 20 minutes, or until brown. Turn the oven down to 100°C, 200°F, gas ¼ and bake for a further 25 to 30 minutes. Cover with foil if the pastry is getting too brown.

To freeze cooked: allow to cool, then wrap and freeze.

To serve after freezing cooked: transfer straight from the freezer to a hot oven (220°C, 425°F, gas 7). After about 10 minutes lower the heat to 180°C, 350°F, gas 4, and continue to cook for a further 30 to 40 minutes, until the pasty has warmed through. Cover with foil if the pastry is getting too brown.

Variations: the recipe given here can be varied by adding thinly sliced turnips or swedes, or chopped parsley. Or – but only if you are not going to freeze the pasties – hard-boiled eggs.

Carrot Quiche

Makes 2 quiches, each serving 4

A recipe from Lorraine, where quiches are said to have been invented – as early as 1586 they were a favourite dish of Charles III, Duke of Lorraine. It seems that the word itself first appeared in 1805, in a book which spoke of 'the antiquity of the quiche of Nancy'. The recipe given below is simple but unusual.

10 oz (275 g) shortcrust pastry	½ pt (300 ml) milk
6 medium carrots	4 eggs
2 medium onions	about ¼ pt (150 ml) double cream
3 oz (75 g) butter	salt and freshly ground pepper
1 oz (25 g) plain flour	freshly grated nutmeg

Heat the oven to 200°C, 400°F, gas 6.

Roll out the pastry and use to line 2 × 7½-in (19-cm) floured flan tins.

Grate or shred the carrots and onions and cook them gently in 2 oz (50 g) of the butter until soft but not brown.

Melt the remaining butter in a small saucepan, add the flour and stir over a low heat for 2 to 3 minutes. Gradually stir in the milk until the sauce is smooth and thick. Remove from the heat.

Beat the eggs and mix with the vegetables and the sauce. Add the cream and season with salt, pepper and freshly grated nutmeg.

Divide between the pastry cases and bake in the centre of the oven for 20 to 30 minutes, until the filling is firm and golden.

To serve immediately: serve hot.

To freeze: allow to cool, then wrap and freeze.

To serve after freezing: transfer straight from the freezer to a hot oven (220°C, 425°F, gas 7). Bake for 20 minutes, then turn the oven down to 190°C, 375°F, gas 5, and bake for a further 20 to 30 minutes. Serve hot.

Smoked Salmon Quiche

Makes 2 quiches, each serving 4–5

This quiche is a good way of using smoked salmon trimmings, which can sometimes be bought quite cheaply – or you may have some left over from a side of smoked salmon. Eat one quiche and freeze the other.

10 oz (275 g) shortcrust pastry
3 oz (75 g) grated gruyère cheese
8 oz (225 g) smoked salmon
  trimmings
½ pt (300 ml) milk
2 oz (50 g) arrowroot or cornflour

4 large eggs
¼ pt (150 ml) double or
  whipping cream
a little freshly grated nutmeg
freshly ground pepper

Heat the oven to 180°C, 350°F, gas 4.

Roll out the pastry and use to line 2 × 7½-in (19-cm) floured flan tins. Spread the cheese over the bottom of the pastry cases. Cut the smoked salmon into thin strips and divide between the cases.

Blend the milk with the arrowroot or cornflour and add the beaten eggs and the cream. Season lightly with nutmeg and pepper and pour over the smoked salmon.

Bake for about 40 minutes, until the filling is risen and golden-brown.

To serve immediately: serve hot.

To freeze: allow to cool, then wrap and freeze.

To serve after freezing: the quiche can be warmed straight from the freezer. Place in a hot oven (220°C, 425°F, gas 7) for 20 minutes, then lower the heat to 180°C, 350°F, gas 4 for a further 30 minutes or so. Cover with a piece of foil if it is getting too brown. If you have thawed the quiche before reheating, it will need only about 20 minutes in a moderate oven (180°C, 350°F, gas 4). Serve warm.

Bakewell Tart

Makes 2 tarts, each serving 4–5

10 oz (275 g) shortcrust pastry
raspberry or apricot jam
4 oz (100 g) butter
4 oz (100 g) sugar

4 oz (100 g) fresh white breadcrumbs
6 oz (175 g) ground almonds
2 eggs
a few drops almond essence

Heat the oven to 180°C, 350°F, gas 4.

Roll out the pastry and use to line 2 × 7½-in (19-cm) floured flan tins. Spread with the jam.

Cream the butter with the sugar. Mix together the breadcrumbs and ground almonds. Gradually add this mixture alternately with the beaten eggs to the creamed mixture, beating well with each addition. Stir in a very few drops of almond essence, but be careful not to add too much, as it is easy to drown the delicate flavour of the filling.

Divide between the pastry cases and bake in the centre of the oven for about 30 minutes, until the filling is firm and the top is golden.

To serve immediately: serve warm or cold with cream.

To freeze: allow to cool, then wrap and freeze.

To serve after freezing: allow to thaw at room temperature and serve as above. If you want to crisp up the pastry a little, transfer straight from the freezer to a moderately hot oven (200°C, 400°F, gas 6) for about 30 minutes.

Almond Tart

Makes 2 tarts, each serving 4–5

A lovely old-fashioned recipe.

10 oz (275 g) shortcrust pastry
4 eggs
7 oz (200 g) sugar
8 oz (225 g) ground almonds

4 tbls (60 ml) whipping or single cream

a little caster sugar

Heat the oven to 200°C, 400°F, gas 6.

Roll out the pastry and use to line 2 × 7½-in (19-cm) floured flan tins.

Mix together the beaten eggs, sugar, ground almonds and cream. Divide the mixture (which should be the consistency of thick cream) between the flan tins and bake for about 15 minutes.

To serve immediately: 2 minutes before the tarts are ready, sprinkle them with a little caster sugar. Serve hot or cold, with cream.

To freeze: allow to cool, then wrap and freeze.

To serve after freezing: transfer straight from the freezer to a moderately hot oven (200°C, 400°F, gas 6) and warm through for about 30 minutes, sprinkling the top with a little caster sugar 2 minutes before you take the tarts out of the oven. Serve hot or cold, with cream.

Note: the almond taste can be accentuated by adding a very few drops of almond essence, but the filling is nicer, and more delicate, without this addition.

Chestnut Tart

Makes 2 tarts, each serving 4–5

10 oz (275 g) shortcrust pastry
8 oz (225 g) chestnuts
about ½ pt (300 ml) half milk and
  half water
3 eggs

4 oz (100 g) sugar
a few drops vanilla essence
4 tbls (60 ml) double cream

sweetened whipped cream

Roll out the pastry and use to line 2 × 7½-in (19-cm) floured flan tins.

To prepare the chestnuts, make a cross in the shells and cook in plenty of boiling water for about 10 minutes. Take them out three or so at a time, and when cool enough to handle peel off the shells and the brown skin. Simmer the chestnuts gently in the milk and water for 20 to 30 minutes until soft. Purée.

Heat the oven to 200°C, 400°F, gas 6.

Separate the eggs. Beat the yolks with the sugar and stir in the chestnut purée, a few drops of vanilla essence and the lightly beaten cream. Whisk the egg whites until they stand in stiff peaks and fold in.

Bake in the centre of the oven for about 40 minutes, until the filling is risen and golden-brown.

To serve immediately: serve warm or cold with lots of lightly beaten, slightly sweetened cream.

To freeze: allow to cool, then wrap and freeze.

To serve after freezing: thaw at room temperature for 4 to 6 hours. If you want to crisp up the pastry a little, transfer straight from the freezer to a moderately hot oven (200°C, 400°F, gas 6) for about 30 minutes.

Blue Cheese Puffs

Serves 6

These are best of all made with roquefort, but they can also be made with stilton or any other rather sharp, dry, blue cheese. They are excellent served hot with drinks, or as a first course with a salad garnish.

1 × 13-oz (375-g) packet puff pastry	2 tbls (30 ml) double cream
8 oz (225 g) roquefort or other blue cheese	2 oz (50 g) walnuts
1 tbls (15 ml) brandy	1 small egg
	1 tbls (15 ml) water

Roll out the pastry ⅛ in (3 mm) thick. Cut into 3-in (7.5-cm) squares.

Crumble the cheese and blend lightly with the brandy, cream and finely chopped walnuts. There is no need to form a smooth paste.

Put a spoonful of the cheese mixture into the centre of each pastry square. Moisten the edges and fold over into a triangle. Seal the edges well and crimp them together. Brush with the egg lightly beaten with water.

To serve immediately: place the triangles on a baking sheet, and bake for 15 to 20 minutes in a hot oven (220°C, 425°F, gas 7) until puffed up and golden-brown. Serve hot.

To freeze: open-freeze, then store in the freezer in an airtight tin or box.

To serve after freezing: transfer straight from the freezer to a hot oven (220°C, 425°F, gas 7) and bake for 40 to 50 minutes. You may need to turn the oven down to about 190°C, 375°F, gas 5 for the last 10 to 15 minutes.

CHOUX PASTRY

Choux pastry is one of a cook's best con tricks. It is extremely simple to make – indeed it is virtually impossible to go wrong – and it has an infinite variety of uses, savoury as well as sweet. It can be suspended in frozen animation at various stages and yet it always presents itself at the table with the air of having been produced by a master chef.

Once you have acquired the art of choux pastry you can use it in any one of its many guises, and you will probably always want to keep some puff shells, baked or unbaked, in the freezer against unexpected contingencies.

Choux Pastry (Basic Mixture)

½ pt (300 ml) water	4½ oz (115 g) plain flour
pinch of salt	4 eggs
bare 4 oz (100 g) butter	

Put the water and salt into a large, heavy saucepan, add the butter cut into small pieces and bring slowly to the boil. Stir until the butter has melted.

Sift the flour, then tip the whole lot, all at once, into the saucepan. Lower the heat, and stir with a wooden spoon until the mixture forms one thick, smooth mass, and leaves the side of the pan clean.

Remove the pan from the heat and beat in the eggs one by one, using a wooden spoon. Do not add the next egg until the previous one has been completely absorbed. You will end up with a smooth glossy paste, buttercup yellow, lukewarm and stiff enough to hold its shape.

Proceed as in individual recipes.

You can also freeze the mixture at this stage if you wish, in polythene bags or in waxed or plastic containers. When you come to use the pastry allow to thaw for several hours, then reheat gently in a saucepan until lukewarm. Proceed as in individual recipes.

Savoury Puffs

Small savoury puffs make excellent cocktail party snacks; larger ones can be served as starters for a meal.

Make the basic choux pastry mixture as above.

Using a piping bag with a 1-in (2.5-cm) nozzle, or two teaspoons, drop small mounds of the mixture on to dampened baking sheets. They should be approximately 1 in (2.5 cm) in diameter for small puffs, or 2½ in (6 cm) in diameter for larger ones (use two dessertspoons for shaping large puffs).

Heat the oven to 220°C, 425°F, gas 7, and bake for 20 to 25 minutes until the puffs have risen and are pale brown.

Remove from the oven and turn the oven off.

Using a sharp, pointed knife, make an incision along the side of each puff, to allow the steam to escape. Put the puffs back on the baking sheets, cut side up, and return to the oven to dry for about 10 minutes, leaving the oven door open. Then remove from the oven and leave to cool on a wire rack.

The puffs can be frozen uncooked if you wish. Open-freeze, then store in rigid containers in the freezer. When you want to cook them, place them in a hot oven (220°C, 425°F, gas 7) and bake for 25 to 30 minutes. Remove from the oven, turn the oven off, and continue as above.

You can also freeze the puffs after they have been baked and before they are filled. Open-freeze, then store in rigid containers in the freezer. When ready to use transfer straight from the freezer to a hot oven (220°C, 425°F, gas 7) and bake for 5 minutes to crisp them, then remove from the oven and leave to cool on a wire rack before filling.

Fill with any of the fillings suggested below.

Fillings for Savoury Puffs

Savoury puffs may be filled with a fine white sauce to which may be added chopped ham, sautéed mushrooms, chopped prawns, shrimps, or smoked salmon.

Sauce

1 oz (25 g) butter
1 oz (25 g) plain flour
¼ pt (150 ml) white wine
¼ pt (150 ml) milk
salt and freshly ground pepper

Melt the butter in a heavy saucepan, add the flour and cook, stirring, without allowing to brown, for 3 minutes.

Add the wine, stir until smooth and cook for 2 minutes, then add the milk and stir until smooth. Season. Bring briefly to the boil, remove from the heat and add the other chosen ingredients.

Allow the sauce to cool and thicken slightly, then fill the puffs, using a teaspoon, and heat through before serving.

The puffs can also be filled with Prawns in Cream and Brandy Sauce (see p. 50) or Chicken Liver Pâté (see p. 25). Serve warm.

Cheese Puffs

These are particularly good served with cocktails or pre-dinner drinks.

Make savoury puffs as above but add 4 oz (100 g) grated gruyère, parmesan or cheddar cheese, or 2 oz (50 g) of a white goat's cheese, lightly mashed, to the basic mixture while it is still warm, immediately after beating in the eggs.

Brush the puffs with a little beaten egg and sprinkle with a little finely grated cheese before baking.

As a variation on the filling, make a sauce as on p. 177 but add 2 oz (50 g) grated cheese (or mashed goat's cheese) after adding the milk, and, if you like, fold in 2 oz (50 g) toasted flaked almonds before filling the puffs.

Gougère

Serves 4–6

A gougère, or large cheese ring, makes an excellent light lunch or supper dish. It can be served plain, with a salad, or you can fill the centre, before serving, with a hot vegetable – asparagus, broad beans, or broccoli, for example – topped with a cheese or hollandaise sauce.

Make the basic mixture, as on p. 175, but use ½ pt (300 ml) milk instead of the ½ pt (300 ml) water. You also need 4 oz (100 g) gruyère or sharp cheddar cheese, cut into small cubes.

When the basic mixture is still warm stir in all but 1 tbls (15 ml) of the cheese.

Heat the oven to 190°C, 375°F, gas 5.

Spoon the mixture into a wetted spring-form tin with a central funnel or a ring mould. Scatter the remaining cheese on top and bake for 35 to 40 minutes, until the top has risen and is golden-brown and firm to the touch.

Serve as suggested above, very hot.

Note: an unusual, but very delicious gougère, can also be made with stilton (a good way of using up stilton after Christmas if it has become a little dry).

Sweet Puffs

Make the basic choux pastry mixture, adding 1 tsp (5 ml) sugar at the same time as the salt (see p. 175).

Large Cream Puffs

The simplest way of serving large sweet puffs is to fill them when cold with sweetened, vanilla-flavoured whipped cream and sprinkle them with sifted icing sugar.

Allow ½ pt (300 ml) double or whipping cream for the quantity given on p. 175, and whip it fairly lightly, so that it is not too stiff.

When baking large puffs to be served in this way, it is nice to mark a light criss-cross on each puff before baking. As they rise, this expands to give the traditional chequered cream puff look.

Sweet puffs can be frozen before or after baking, in the same way as savoury puffs. They may also be frozen after filling, though they will take 3 to 4 hours at room temperature to thaw, and they will not be quite as light as when they have been filled just before serving.

Éclairs

Make finger-sized éclairs for dainty tea parties, and rather larger ones to serve for dessert.

Make the sweetened basic choux pastry mixture (see p. 178) and either use éclair tins or pipe into the characteristic long thin shape, slightly stubby at each end, by using a piping bag with a plain ½-in (1-cm) nozzle.

In addition you will need:

Filling

½ pt (300 ml) double or whipping cream   1 tsp (5 ml) icing sugar

Icing

4 tbls (60 ml) water   1 tsp (5 ml) instant coffee powder
1 tbls (15 ml) sugar   3 oz (75 g) icing sugar
2 oz (50 g) plain or bitter chocolate   ½ tsp (2.5 ml) corn or vegetable oil

Fill the éclairs when baked and cooled (as for Savoury Puffs, see p. 176) with the sweetened whipped cream.

To make the icing, put the water and sugar in a small saucepan and bring slowly to the boil. Stir until the sugar has dissolved, then boil briskly for 1 minute. Remove from the heat and add the chocolate, broken into pieces. Stir until melted. Beat in the coffee and icing sugar until smooth, then stir in the oil.

Using a knife, coat the top of each éclair thickly with this icing, then leave to set before serving.

To freeze: finished éclairs can be frozen. Open-freeze, then store in rigid containers.

To serve after freezing: leave the éclairs at room temperature for at least 1 hour before serving.

Profiteroles or Small Cream Puffs

Using the sweetened basic mixture, make small puffs as described on p. 176.

In addition, you will need:

Filling
½ pt (300 ml) double or
   whipping cream
1 tsp (5 ml) icing sugar

or
½ pt (300 ml) vanilla ice cream

Sauce
5 oz (150 g) sugar
¼ pt (150 ml) water
1 tbls (15 ml) cocoa powder

2 oz (50 g) plain or bitter dessert
   chocolate

When the puffs are cool, fill with the lightly whipped sweetened cream or with ice-cream.

To serve immediately: make the sauce. Bring the sugar, water and cocoa to the boil and boil fiercely for 5 minutes. Remove from the heat, add the chocolate broken into pieces and stir until it has melted and the sauce is quite smooth. Bring briefly to the boil again.

Pile the profiteroles into a dish and pour over the hot sauce. Serve at once.

To freeze: Open-freeze, then store in rigid containers.

To serve after freezing: thaw for at least 30 minutes at room temperature before serving as above.

Croquembouches

Serves 6–8

A most impressive sweet, which forms a lovely centrepiece for a buffet table but is quite simple to make. It is best to make and bake the puffs well ahead of the party and store them in the freezer, and then to fill and assemble them on the day itself.

Make small sweet puffs as on p. 178.

In addition you will need:

Filling

½ pt (300 ml) double or whipping cream	1 tbls (15 ml) Grand Marnier or orange curaçao
1 tsp (5 ml) icing sugar	finely grated rind of 1 orange

Topping

5 oz (150 g) sugar	finely grated rind of ½ orange
2 tbls (30 ml) water	

Whip the cream lightly with the icing sugar, and when it begins to thicken slowly whisk in the liqueur and the orange rind. Fill each puff with this cream.

Melt the sugar in a heavy saucepan with the water and orange rind. When it begins to caramelize and turn golden-brown, remove from the heat and either dip in each puff briefly, or pour a little of the mixture over each one. You must work quickly so that the mixture does not set. If it does harden before you have finished coating the puffs, heat it again very gently.

The puffs can be arranged on a large serving platter, allowing 3 to 4 per person. For special occasions, make a double quantity of topping. Dip in each puff after it has been filled, and construct a pyramid shape on a cake stand, using a little topping to cement the puffs together. Pour the remaining topping slowly over the finished edifice, so that it forms drips and runnels. Leave to set. You can then decorate with small cake candles, which, when lit, will make the dessert look quite sensational.

FILA OR STRUDEL PASTRY

Fila are the whisper-thin pastry sheets much used in Greek and Middle Eastern cooking, traditionally so fine that you can read a newspaper through them. They can be bought ready-made from Greek shops and most good delicatessens in packs of approximately twenty-four sheets. German or Austrian strudel pastry can be used in the same way.

Although the pastry sheets will only remain good in the refrigerator for up to a week, they can be kept for several months in the freezer. As this pastry lends itself to numerous savoury and sweet dishes, it is useful to keep a packet in reserve.

Once the pastry is exposed to air, it dries up very quickly, so it must be kept damp and pliable while working with it. When you have unwrapped it, peel off one sheet at a time, and keep a damp cloth over the remainder. Use up a whole packet at a time (making several dishes at once, if you like), as the sheets will not keep well once the packet has been opened.

Crispy Cheese Pie

Serves 6

10–12 sheets fila pastry	2 eggs
2 oz (50 g) butter	1 tbls (15 ml) finely chopped mint or
2 medium onions	1½ tsp (7.5 ml) finely chopped
1 clove garlic	marjoram
1 tbls (15 ml) oil	salt and freshly ground pepper
8 oz (225 g) curd cheese	pinch of freshly grated nutmeg
8 oz (225 g) cream cheese	

Frozen fila pastry must be allowed to thaw completely as otherwise it will break.

Heat the oven to 220°C, 425°F, gas 7.

Melt the butter and brush the inside of a 10 × 12-in (25 × 30-cm) baking tin. Lay in the first fila sheet, allowing it to overhang the edges, and brush lightly all over with melted butter. Cover with the next sheet, brush with butter and continue until 5 or 6 sheets have been used.

Chop the onions and garlic together quite finely and fry in the oil until soft and golden-brown. Drain well.

Mix the cheeses together with a fork, and mix in the lightly beaten eggs. Add the onion and garlic, the herbs, seasoning and nutmeg.

Spread this mixture evenly over the pastry and cover with the remaining fila sheets, brushing each with melted butter. Fold over the edges to seal.

To serve immediately: brush the top of the pie with butter and bake in the oven for 35 to 40 minutes until the pie is risen and golden-brown on top.

To freeze: wrap the unbaked pie in clingfilm, being careful to exclude all the air, then overwrap with foil.

To cook after freezing: remove the wrapping and transfer straight from the freezer to a hot oven (220°C, 425°F, gas 7), brushing with a little melted butter first. Bake for 1 to 1¼ hours, lowering the oven temperature towards the end if the top is becoming too brown.

Note: this pie can also be made with other fillings:

Meat Filling

1 tbls (15 ml) oil	½ tsp (2.5 ml) sugar
2 large onions	½ tsp (2.5 ml) ground allspice
1 clove garlic	½ tsp (2.5 ml) ground cinnamon
1 lb (500 g) minced lamb or beef, or a mixture of the two	salt and freshly ground pepper
1 small can concentrated tomato purée	

Heat the oil in a large frying pan and fry the finely chopped onions and garlic until golden and transparent. Add the meat and stir until it has all changed colour. Add the remaining ingredients, stir well and season.

Spinach Filling

2 lb (1 kg) fresh spinach or 1 lb (500 g) frozen	good pinch of freshly grated nutmeg
2 oz (50 g) butter	4 oz (100 g) gruyère, cheddar, Greek feta or cottage cheese
salt and freshly ground pepper	

Blanch fresh spinach in boiling salted water for 1 minute, drain well and leave until cool enough to handle. Squeeze out all the moisture with your hands. Chop the spinach roughly and cook in the butter until just tender. Frozen spinach should be heated through gently from frozen. When thawed, raise the heat to evaporate all moisture, then add the butter and cook until just tender. Season, add the nutmeg and the grated or lightly mashed cheese and mix well before filling.

Spinach 'Bricks'

These delicious, crisp little parcels can be made in different sizes – tiny ones for handing round with drinks or medium-sized for a first course at dinner, while quite largish ones will make a good supper dish.

1 lb (500 g) fila or strudel pastry

1 lb (500 g) frozen spinach

2 oz (50 g) butter

8 oz (225 g) feta, ricotta or
  curd cheese

1 egg

salt and freshly ground pepper

freshly grated nutmeg

4 oz (100 g) cooked ham (optional)

oil for deep frying

Allow frozen fila pastry to thaw completely as otherwise it will break. Cook the spinach from frozen with a scant tablespoon of water until just tender. Mix in half the butter, the cheese (grated if you are using feta) and the beaten egg and season well, especially with nutmeg. Add the finely chopped ham if you are using it. Leave to cool.

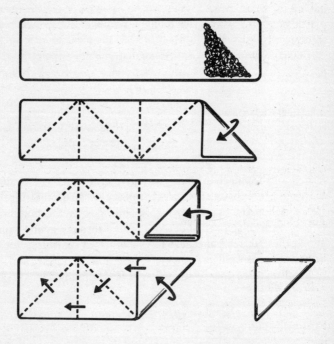

Heat the remaining butter until just melted. Separate the fila sheets. Brush one sheet with melted butter and cut into long strips, 2 in (5 cm),

3 in (7.5 cm) or 5 in (13 cm) wide, depending on what size you want the finished 'bricks' to be. Place 1 tsp (5 ml) (2 tsp/10 ml for the largest size) of the spinach mixture near the top of the first strip, and fold the pastry over it in a triangle, so that the top edge lies over the side edge. Fold this triangle over again, so that the top is now straight again. Fold the triangle across to the other side, and continue folding in this way until the strip is used up. By this method all the sides of the triangle should have been closed at least once. Continue with the remaining fila sheets.

Deep-fry the 'bricks', a few at a time, in hot oil until golden-brown. Drain and serve very hot.

To freeze: drain on kitchen paper and leave to cool, then spread out on baking trays and freeze. When frozen hard, place in polythene bags or plastic boxes (they are quite fragile even when frozen) and store in the freezer.

To serve after freezing: transfer straight from the freezer to a hot oven (220°C, 425°F, gas 7), for 10 to 20 minutes, depending on size.

Baklava

Serves 4–6

A delicious, very sweet Middle Eastern dessert or sweetmeat.

12 sheets fila pastry	1 tbls (15 ml) sugar or honey
2 oz (50 g) melted butter	
4 oz (100 g) unblanched almonds or hazelnuts or a mixture of the two	

Syrup

¼ pt (150 ml) water	1 tbls (15 ml) rosewater or orange
3 tbls (45 ml) sugar or honey	blossom water
good squeeze of lemon juice	

Frozen fila pastry must be allowed to thaw completely as otherwise it will break.

Heat the oven to 190°C, 375°F, gas 5.

Fit 6 fila sheets, each brushed with melted butter, into a 7 × 11-in (18 × 28-cm) baking tin. Chop the nuts quite coarsely, mix with the sugar or honey and spread over the pastry. Cover with the remaining pastry sheets, brushing each one with butter.

Mark the top into lozenge shapes with a sharp knife, and bake in the oven for 35 to 40 minutes until puffed up and golden, raising the heat a little towards the end if necessary.

To serve immediately: while the baklava is baking, bring the sugar or honey and water to the boil together with the lemon juice. Simmer until it thickens to a light syrup. Refrigerate. As soon as you have removed the baklava from the oven, add the rose- or orange blossom water to the cooled syrup and pour over the hot baklava. Cut into lozenge shapes and serve hot or cold.

To freeze: leave to cool, then wrap well and freeze.

To serve after freezing: transfer straight from the freezer to a moderately hot oven (190°C, 375°F, gas 5) and bake for 15 to 20 minutes, until heated through. Meanwhile make the syrup and serve as above.

Cakes

✳✳✳✳✳✳✳✳✳✳✳✳✳✳✳✳✳✳✳✳

Three cheers for the freezer when it comes to making cakes, for generally it is just as quick to double the quantities, make two cakes, eat one straight away and freeze the second.

Most cakes – with the exception of creamy ones – freeze excellently. In general there is less point in freezing rich fruit cakes, which keep well, and indeed improve, stored in airtight tins. But most other kinds, especially sponges, which go dry quickly, or fresh fruit and vegetable cakes, such as carrot or marrow, which may start to grow whiskers if they are stored in a tin for too long, are perfect subjects for freezing.

Cakes should be frozen, well wrapped, the moment they are cool, and should be thawed, freed from their freezer wrappings, at room temperature.

Most cakes can be frozen iced, though some are better iced after they have come out of the freezer. They should be open-frozen, so that the icing is not damaged, then wrapped and replaced in the freezer. When they are taken out of the freezer the wrapping should be removed, so that as the icing thaws it will not be spoilt by sticking to the wrapping.

Carrot Cake

Makes 2 × 1-lb (450-g) cakes

This moist, wholesome loaf, which can be iced and eaten as a cake or sliced and buttered like sweet bread, keeps particularly well in the freezer.

1 lb (500 g) carrots
finely grated rind and juice of
   1 lemon or orange
6 oz (175 g) butter or margarine
12 oz (350 g) soft brown sugar or
   honey
4 eggs
1 tsp (5 ml) vanilla essence

8 oz (225 g) plain flour
8 oz (225 g) wholemeal flour
1 tsp (5 ml) salt
2 tsp (10 ml) baking powder
2 tsp (10 ml) ground cinnamon
pinch of ground allspice
4 oz (100 g) hazelnuts, walnuts or
   pecans (optional)

Icing (optional)
4 oz (100 g) butter
4 oz (100 g) cream cheese

2 tbls (30 ml) honey or icing sugar
a little grated lemon or orange rind

Heat the oven to 180°C, 350°F, gas 4.

Butter 2 × 1-lb (450-g) loaf tins.

Grate the carrots and sprinkle on the lemon or orange juice. Cream the butter or margarine with the sugar or honey until light and fluffy, beat in the eggs one by one and add the vanilla essence and the lemon or orange rind.

Sift the flours, salt, baking powder and spices together at least twice, until light but well blended.

Add the dry ingredients alternately with the carrots to the creamed mixture and fold in lightly but thoroughly with a spoon or spatula. Fold in the roughly chopped nuts if you are using them.

Pour into the prepared tins and bake in the oven for 1 hour, or until a skewer inserted into the centre comes out clean. Remove from the oven and leave to cool in the tins for 15 minutes, then turn out on to a wire rack and leave to cool completely.

To serve immediately: beat the icing ingredients together until very light and fluffy, then spread thickly over the top of the loaves.

To freeze: wrap well in clingfilm or foil, then freeze.

To serve after freezing: unwrap and leave to thaw at room temperature for 3 to 4 hours. Ice as above.

Sue's Zucchini Cake

Makes 2 × 1-lb (450-g) cakes

A recipe from Canada. Like the carrot cake, it has the merit of keeping well, and of having a pleasantly moist texture and an unusual taste. It is a good way of using up those zucchini (courgettes) which will insist on growing too fast.

8 oz (225 g) zucchini (courgettes)	2 tsp (10 ml) ground cinnamon
5 oz (150 g) butter or margarine	3 oz (75 g) seedless raisins (optional)
12 oz (350 g) sugar	2 oz (50 g) chopped nuts
4 eggs	2 tsp (10 ml) vanilla essence
11 oz (300 g) self-raising flour	

Heat the oven to 180°C, 350°F, gas 4.

Butter 2 × 1-lb (450-g) loaf tins.

Grate the unpeeled zucchini and set aside.

Cream the butter or margarine with the sugar until light and fluffy. Beat the eggs well and add alternately with the flour, beating in a little at a time and making sure that each is well mixed in before the next addition. Beat in the cinnamon and the zucchini, and finally the raisins, if you are using them, the chopped nuts and vanilla essence.

Divide equally between the prepared tins, and bake in the centre of the oven for about 1 hour or until a skewer inserted into the centre comes out clean. Turn the cakes out on to a wire rack and leave to cool completely.

To freeze: wrap well in clingfilm or foil, then freeze.

To serve after freezing: unwrap and leave to thaw at room temperature for 3 to 4 hours. Although delicious if eaten as soon as it has thawed, this cake is even better if left to mature in a cake tin for 2 to 3 days, but should then be eaten fairly quickly.

Note: the icing for Carrot Cake (see previous recipe) is also good with this Zucchini Cake.

Spice Cake

Makes 1 × 8-in (20-cm) cake

This spicy loaf can be eaten buttered or plain, as you prefer.

3 oz (75 g) lard
3½ oz (90 g) barbados sugar
½ cup (100 ml) black treacle
½ tsp (2.5 ml) vanilla essence
2 eggs
10 oz (275 g) plain flour
2 tsp (10 ml) baking powder

½ tsp (2.5 ml) bicarbonate of soda
½ tsp (2.5 ml) salt
1 tsp (5 ml) ground ginger
1 tsp (5 ml) ground cinnamon
good pinch of ground cloves
good pinch of freshly grated nutmeg
¼ pt (150 ml) boiling water

Heat the oven to 160°C, 325°F, gas 3.

Line an 8-in (20-cm) cake tin with greased greaseproof paper.

Beat the lard with the sugar until thick and creamy. Stir in the treacle
and the vanilla essence. Add the eggs one at a time, beating vigorously
with each addition. Sift the remaining dry ingredients thoroughly
together and beat them in gradually, alternately with the boiling water.

Bake in the centre of the oven for 45 minutes, or until a skewer
inserted into the centre comes out clean. Turn out on to a wire rack and
leave to cool completely.

To freeze: as soon as the cake is cool, wrap, then freeze.

To serve after freezing: unwrap and thaw at room temperature for 4 to
6 hours.

Hiker's Cake

Makes 2 × 8-in (20-cm) cakes

So called because it is nutritious enough to sustain a hungry hiker or
sportsman through a long, damp day, and solid enough for a slice to keep
its shape in his pocket. It is a marvellous stand-by to have in the freezer,
where it will keep virtually indefinitely.

1 lb (450 g) butter
1 lb (450 g) soft dark brown sugar
1 lb 4 oz (550 g) plain flour
1 tbls (15 ml) ground mixed spice
8 eggs
8 oz (225 g) walnuts or pecan nuts

8 oz (225 g) mixed peel
1 lb (450 g) seedless raisins
1 lb (450 g) sultanas
about ½ pt (300 ml) Guinness or
     other beer or stout

Heat the oven to 160°C, 325°F, gas 3.

Line 2 × 8-in (20-cm) cake tins with greased greaseproof paper.

Cream the butter with the sugar until light and fluffy. Sift the flour with the spice. Beat the eggs one by one into the creamed mixture, then gently fold in the flour.

Roughly chop the nuts and the peel and add, together with the dried fruit. Add approximately half the beer to the mixture, enough to give it a soft dropping consistency.

Divide the mixture between the two tins. Bake in the oven for 1 hour, then lower the heat to 150°C, 300°F, gas 2 and bake for a further 1½ hours, or until a skewer inserted into the centre comes out clean.

Remove from the oven and allow to cool for about 1 hour before turning out of the tins.

When the cakes are quite cool, turn them upside down on to wire racks placed over a plate or dish, prick well with a long fork or skewer and gently trickle on the remaining beer.

The cake is best left to mature for a few days before eating.

To freeze: when the cake has absorbed all the beer, wrap well in clingfilm, then with foil, and freeze.

To serve after freezing: it is best to remove the cake from the freezer at least one day before serving, but if this is not possible, allow at least 5 to 6 hours at room temperature to thaw.

Coffee Cake

Makes 2 × 8-in (20-cm) cakes

Cake

8 oz (225 g) butter or margarine
8 oz (225 g) sugar
4 eggs

8 oz (225 g) self-raising flour
2 tsp (10 ml) instant coffee powder
2 tbls (30 ml) boiling water

Filling

1 oz (25 g) softened butter
6 oz (175 g) icing sugar
2 tsp (10 ml) instant coffee powder

1 tbls (15 ml) boiling water

Icing

6 oz (175 g) icing sugar
2 tsp (10 ml) instant coffee powder
1 tbls (15 ml) boiling water

walnut halves (optional)

Heat the oven to 180°C, 350°F, gas 4.

To make the cake, line 2 × 8-in (20-cm) sponge flan tins with greased greaseproof paper.

Cream the butter or margarine with the sugar until light and fluffy. Add the well-beaten eggs alternately with the sifted flour, making sure that each addition is thoroughly mixed in before making the next. Finally fold in the instant coffee dissolved in the boiling water.

Beat the mixture well, then divide equally between the prepared tins. Bake in the centre of the oven for about 35 minutes, or until the cakes are just firm to the touch and a skewer inserted into the centre comes out clean.

Turn out on to a wire rack and leave to cool completely, then fill and ice.

To make the filling, cream the butter well with the icing sugar, then stir in the instant coffee dissolved in the boiling water. Cut each cake in half horizontally and sandwich together with the filling.

To make the icing, mix thoroughly the icing sugar and the instant coffee dissolved in the boiling water. Spread evenly over the cakes.

To serve immediately: decorate with walnut halves, if you are using them, before the icing sets.

To freeze: open-freeze, then wrap and replace in the freezer.

To serve after freezing: thaw at room temperature for 3 to 4 hours. Decorate as above.

Cut-and-Come-Again Cake

Makes 1 × 7 × 12-in (18 × 30-cm) cake

An excellent stand-by. It freezes beautifully and is marvellous for hungry children, picnics, car journeys, holidays or what you will. The quantity of dried fruit can be varied, according to how rich you want the cake to be. Eat it as it is, or spread with butter. The quantity given below makes a large and substantial cake, so you may like to freeze half (one large cake seems to turn out better than two small ones with this recipe).

8 oz (225 g) butter or margarine
1¼ lb (550 g) self-raising flour
8 oz (225 g) sugar
3–4 eggs
12 oz (350 g) mixed dried fruit
(or more if you like)

a few glacé cherries
about 2 oz (50 g) walnut pieces
(optional)
a little milk

Heat the oven to 180°C, 350°F, gas 4.

Line a 7 × 12-in (18 × 30-cm) tin about 1½–2 in (4–5 cm) deep with greased grease proof paper.

Rub the butter or margarine into the flour and add the sugar. Beat the eggs and mix in well. Stir in the dried fruit. Cut the glacé cherries and walnuts, if you are using them, into smallish pieces and stir in. Add enough milk to make a stiff, 'rock-cake-like' consistency. Beat the mixture well.

Turn into the prepared tin and bake in the oven for 30 to 40 minutes, until the cake has risen and is turning golden, then lower the heat to 120°C, 250°F, gas ½, and bake for a further 35 to 40 minutes, until a skewer inserted into the centre comes out clean.

Turn out on to a wire rack and leave to cool completely.

To freeze: wrap, then freeze.

To serve after freezing: unwrap and thaw at room temperature for 4 to 6 hours.

Orange Snow Cake

Makes 2 × 7-in (18-cm) cakes

This cake keeps beautifully moist, and has an unusual flavour and texture.

12 oz (350 g) butter
finely grated rind of 2 oranges
12 oz (350 g) caster sugar
4 eggs
¼ pt (150 ml) water

1¼ lb (550 g) self-raising flour
5 tbls (75 ml) thick-cut marmalade
4 oz (100 g) candied peel
6 oz (175 g) walnuts

Heat the oven to 180°C, 350°F, gas 4.

Butter and lightly flour 2 cake tins each about 7 in (18 cm) in diameter and 3 in (7 cm) deep.

Cream the butter with the grated orange rind and sugar until light and fluffy. Separate the eggs. Beat the yolks with the water. Add the flour and the yolks alternately to the creamed mixture, beating in each addition well before you add the next. Stir in the marmalade, the finely chopped candied peel and the chopped walnuts. Fold in the stiffly beaten egg whites.

Divide the mixture between the prepared tins and bake in the centre of the oven for 1¼ to 1½ hours, until a skewer inserted in the centre comes out clean.

Turn out on to a wire rack and leave to cool completely.

To freeze: as soon as the cake is cold, wrap, then freeze.

To serve after freezing: unwrap and thaw at room temperature for 4 to 6 hours.

Note: you can if you like finish with a light orange icing. Mix 2–3 tbls (30–45 ml) fresh orange juice with enough sifted icing sugar to make a mixture sufficiently stiff to spread over the top of the cake. Leave for 2 to 3 hours to harden. The cake can be frozen iced; open-freeze, and when the icing is quite hard wrap the cake and return to the freezer.

Honey Cake

Makes 1 × 8 × 12-in (20 × 30-cm) cake

The quantities given below make quite a large cake, so you may want to cut it in two when it is cold and freeze half.

6 oz (175 g) honey
4 oz (100 g) soft light brown sugar
5 oz (150 g) butter
1 tbls (15 ml) water

2 eggs
7 oz (200 g) self-raising flour
1–2 oz (25–50 g) flaked almonds

Heat the oven to 180°C, 350°F, gas 4.

Line an 8 × 12-in (20 × 30-cm) tin about 1¼ in (3 cm) deep with greased greaseproof paper.

Weigh the honey. There are two ways of doing this: you can either weigh the sugar, spread it over the weighing bowl and weigh the honey on this bed, so that you have 10 oz (275 g) of sugar and honey combined; or you can place the honey jar on the scales and remove 6 oz (175 g).

Put the honey, sugar, butter and water into a saucepan and heat gently until the sugar has melted. Transfer to a mixing bowl and add the well-beaten eggs. Stir in the sifted flour.

Pour into the prepared tin and scatter generously with flaked almonds. Bake in the centre of the oven for 30 to 35 minutes, until a skewer inserted into the centre comes out clean.

Turn out on to a wire rack and leave to cool completely. Cut into rectangles to serve.

To freeze: as soon as the cake is cold, wrap, then freeze.

To serve after freezing: unwrap and leave to thaw at room temperature for 4 to 5 hours.

Madeira Cake

Makes 2 × 8-in (20-cm) cakes

A useful all-purpose cake which can be eaten without additions; or it can be filled with jam or with butter icing (made of equal parts of butter and sifted icing sugar beaten together). You can also add glacé cherries and walnuts, or perhaps caraway seeds to make a delicious seed cake that is rarely met with nowadays and is reminiscent of Regency ladies delicately nibbling cake and sipping madeira wine.

Basic mixture
8 oz (225 g) butter or margarine
8 oz (225 g) sugar
10 oz (275 g) self-raising flour

4 eggs
2 tbls (30 ml) milk

Heat the oven to 180°C, 350°F, gas 4.

Line 2 × 8-in (20-cm) sponge flan tins with greased paper.

Cream the butter or margarine with the sugar until light and fluffy. Gradually beat in the flour alternately with the well-beaten eggs, making sure that each addition is well mixed in before adding the next. Stir in the milk.

Divide the mixture equally between the prepared tins. Bake in the centre of the oven for 30 to 35 minutes, or until the cake is slightly firm to the touch and a skewer inserted into the centre comes out clean.

Turn out on to a wire rack and leave to cool completely.

To freeze: as soon as the cake is cold, wrap well, then freeze.

To serve after freezing: unwrap and thaw at room temperature for 3 to 4 hours.

Variations:

Cherry and Walnut Cake: beat into the mixture 6 oz (175 g) quartered glacé cherries and 4 oz (100 g) chopped walnuts. Cook for 5 to 10 minutes longer than the basic mixture.

Seed Cake: add 5–6 tsp (25–30 ml) caraway seeds.

Chocolate Cake: instead of 10 oz (275 g) flour, use only 8 oz (225 g) flour and 2 oz (50 g) sieved drinking chocolate powder or 1 oz (25 g) cocoa powder.

Orange or Lemon Cake: add to the mixture the finely grated rind of 2 oranges or lemons, and sandwich together with butter icing to which you have added a little orange or lemon juice.

Chocolate Cake

Makes 2 × 10-in (25-cm) cakes

No family freezer is complete without at least one chocolate cake against unforeseen emergencies. This is a rich, solid cake which will please adults and children alike, and does equally well for teatime or as a dessert, especially if served with whipped cream or ice-cream. These quantities make 2 large cakes – eat one and freeze one. The cake freezes well after it has been filled and iced.

1 lb (450 g) butter or margarine
1 lb (450 g) granulated sugar, or
   half granulated and half soft
   light brown
6 oz (175 g) cocoa powder
4 oz (100 g) drinking chocolate
   powder

1 lb (450 g) self-raising flour
8 oz (225 g) ground almonds
8 eggs
2 tbls (30 ml) strong coffee
2 tbls (30 ml) sweet sherry

Filling
6 oz (175 g) plain or bitter chocolate
2 tbls (30 ml) strong coffee
1 tbls (15 ml) sweet sherry (optional)

6 oz (175 g) butter
10 oz (275 g) icing sugar

Icing
8 oz (225 g) plain or bitter chocolate
¼ pt (150 ml) water
4 oz (100 g) icing sugar

1 oz (25 g) butter

split blanched almonds (optional)

Heat the oven to 180°C, 350°F, gas 4.

Butter 2 × 10-in (25-cm) round cake tins.

Cream the butter or margarine with the sugar until very light and fluffy. Sift together the cocoa, drinking chocolate, flour and ground almonds. Beat the eggs together lightly in a separate bowl. Gradually add the eggs to the creamed mixture alternately with the dry ingredients, and finally add the coffee and sherry. Do not overbeat.

Divide equally between the prepared tins and bake in the oven for about 1 hour or until a skewer inserted into the centre comes out clean.

Remove from the oven and leave to cool in the tins for 10 to 15 minutes, then turn out on to wire racks and leave to cool completely.

To make the filling, break the chocolate into a small bowl, add the coffee and the sherry, if you are using it, and set over a saucepan of simmering water until the chocolate has melted. Stir until smooth and leave to cool slightly.

Beat the butter with the icing sugar until light and fluffy, then beat in the cooled chocolate mixture.

Cut the cakes in half horizontally and sandwich together with the filling.

To make the icing, break the chocolate into a small saucepan, add the water and icing sugar and bring slowly to the boil, stirring all the time. As soon as the mixture begins to thicken and large bubbles appear, remove from the heat, stir in the butter and pour quickly over the cakes, smoothing the icing on if necessary with a wet palette knife. (If the icing sets too quickly, heat again with a little more water.)

Decorate with split almonds, if you are using them, before the icing sets.

To freeze: allow the icing to set and become completely dry, then either freeze in a box or tin or open-freeze and wrap in clingfilm.

To serve after freezing: allow to thaw for 6 to 8 hours at room temperature.

Chocolate Chip Cookies

The essence of a cookie is that it should be freshly baked. This is especially important with these cookies if you use chunks rather than chips of chocolate, since the chocolate will remain soft or even runny on the day of baking but will set the following day. For really irresistible cookies, therefore, it is best to bake them merely hours before serving, but the dough can be made up in quantity and kept in the freezer for weeks at a time.

8 oz (225 g) butter or margarine
4 oz (100 g) granulated sugar
4 oz (100 g) soft light brown sugar
2 large eggs
a few drops of vanilla essence

12 oz (350 g) plain flour
½ tsp (2.5 ml) baking powder
pinch of salt
6 oz (175 g) plain or bitter chocolate
   or chocolate chips

Cream the butter or margarine with the sugars until light and fluffy, then beat in the eggs and vanilla essence.

Sift the flour with the baking powder and salt and beat in gently but thoroughly.

Chop the chocolate into quite large chunks and fold into the mixture, or stir in the chocolate chips.

Refrigerate the mixture for 30 minutes until it is stiff enough to handle, then shape into balls the size of a large walnut.

To serve immediately: Space well apart on a baking sheet. Bake in a moderate oven (180°C, 350°F, gas 4) for 10 to 15 minutes, until the cookies have spread and set in the centre, and have just begun to brown at the edges (they are nicest just a little undercooked). Transfer to wire racks and leave to cool. Serve as soon as possible.

To freeze: store the cookies in polythene bags or plastic boxes.

To cook after freezing: Allow to thaw at room temperature for about 30 minutes and bake as above, or bake straight from the freezer, allowing 20 to 25 minutes' baking time.

Index

****